Understanding How Science Explains the Wor

All people desire to know. We want to not only know what has happened, but also *why* it happened, *how* it happened, whether it *will* happen again, whether it can be *made* to happen or not happen, and so on. In short, what we want are explanations. Asking and answering explanatory questions lies at the very heart of scientific practice. The primary aim of this book is to help readers understand how science explains the world. This book explores the nature and contours of scientific explanation, how such explanations are evaluated, and how they lead to knowledge and understanding. As well as providing an introduction to scientific explanation, it also tackles misconceptions and misunderstandings, while remaining accessible to a general audience with little or no prior philosophical training.

Kevin McCain is Professor of Philosophy at the University of Alabama at Birmingham. His academic research interests lie in epistemology and philosophy of science, focusing on the role of explanatory reasoning in the production of scientific knowledge. He is the author of *The Nature of Scientific Knowledge* (Springer, 2016), co-author of *Uncertainty: How It Makes Science Advance* (Oxford University Press, 2019), and co-editor of *What Is Scientific Knowledge?* (Routledge, 2019).

The ***Understanding Life*** series is for anyone wanting an engaging and concise way into a key biological topic. Offering a multidisciplinary perspective, these accessible guides address common misconceptions and misunderstandings in a thoughtful way to help stimulate debate and encourage a more in-depth understanding. Written by leading thinkers in each field, these books are for anyone wanting an expert overview that will enable clearer thinking on each topic.

Series Editor: Kostas Kampourakis http://kampourakis.com

Published titles:

Understanding Evolution	Kostas Kampourakis	9781108746083
Understanding Coronavirus	Raul Rabadan	9781108826716
Understanding Development	Alessandro Minelli	9781108799232
Understanding Evo-Devo	Wallace Arthur	9781108819466
Understanding Genes	Kostas Kampourakis	9781108812825
Understanding DNA Ancestry	Sheldon Krimsky	9781108816038
Understanding Intelligence	Ken Richardson	9781108940368
Understanding Metaphors in the Life Sciences	Andrew S. Reynolds	9781108940498
Understanding Cancer	Robin Hesketh	9781009005999
Understanding How Science Explains the World	Kevin McCain	9781108995504

Forthcoming:

Understanding Human Metabolism	Keith N. Frayn	9781009108522
Understanding Race	Rob DeSalle and Ian Tattersall	9781009055581
Understanding Fertility	Gab Kovacs	9781009054164
Understanding Human Evolution	Ian Tattersall	9781009101998
Understanding Forensic DNA	Suzanne Bell and John M. Butler	9781009044011
Understanding Natural Selection	Michael Ruse	9781009088329
Understanding Creationism	Glenn Branch	9781108927505
Understanding Species	John S. Wilkins	9781108987196
Understanding the Nature–Nurture Debate	Eric Turkheimer	9781108958165

Understanding How Science Explains the World

KEVIN MCCAIN
University of Alabama at Birmingham

CAMBRIDGE
UNIVERSITY PRESS

University Printing House, Cambridge CB2 8BS, United Kingdom

One Liberty Plaza, 20th Floor, New York, NY 10006, USA

477 Williamstown Road, Port Melbourne, VIC 3207, Australia

314–321, 3rd Floor, Plot 3, Splendor Forum, Jasola District Centre, New Delhi – 110025, India

103 Penang Road, #05–06/07, Visioncrest Commercial, Singapore 238467

Cambridge University Press is part of the University of Cambridge.

It furthers the University's mission by disseminating knowledge in the pursuit of education, learning, and research at the highest international levels of excellence.

www.cambridge.org
Information on this title: www.cambridge.org/9781316518175
DOI: 10.1017/9781108997027

First published 2022

Printed in the United Kingdom by TJ Books Limited, Padstow Cornwall

A catalogue record for this publication is available from the British Library.

ISBN 978-1-316-51817-5 Hardback
ISBN 978-1-108-99550-4 Paperback

"This engaging book effectively introduces a wide range of philosophical ideas about scientific explanation in an accessible way. It's attentive to nuances but avoids getting bogged down in details and debates."

Angela Potochnik, Professor of Philosophy and Director of the Center for Public Engagement with Science, University of Cincinnati

"Kevin McCain's excellent book zooms in on the role of *explanation* in science and links it with scientific *understanding*. McCain has the enviable gift to write a gentle introduction for the novice reader that also provides a fresh perspective that is interesting for the specialist. Overall, this book is an accessible and illuminating contribution to the literature on scientific explanation."

Olaf Dammann, Professor of Public Health and Community Medicine, Tufts University

"In this concise and elegant book, McCain provides a superb overview of current thinking about the nature of explanation in science, correcting common misunderstandings and providing a clearly written, entertaining, and insightful guide to the enterprise of understanding the world."

Michael Strevens, Professor of Philosophy, New York University

"*Understanding How Science Explains the World* is a very impressive achievement. It draws on and develops some of the most important philosophical views on the nature of explanation, while carefully engaging throughout with important examples from the history of science (including quite recent history, which takes into account scientific attempts to explain and understand COVID-19). Highly recommended."

Stephen R. Grimm, Professor of Philosophy, Fordham University

For Declan

Contents

Foreword

Why do birds have wings? Why do offspring resemble their parents? Why did the nonavian dinosaurs become extinct? How do organisms develop? How do species evolve? How do ecosystems remain stable despite perturbations? These and other "why?" and "how?" questions are very common in science – in fact, these are the kinds of questions that scientists usually ask. When the answers to such questions provide accounts of why and how something occurred in the past or regularly occurs, and these accounts in turn provide understanding of the respective phenomena, then these answers are more than that: they are explanations. In this philosophically informed book, Kevin McCain provides a clear and accessible introduction to some core issues in the philosophy of science. He explains what makes explanations different from descriptions, what is the role of causes in explanations, how explanations contribute to understanding, what is the structure of scientific explanations, their potential but also their limits. McCain provides a comprehensive introduction to scientific explanation that is at the same time comprehensible; and these are features that do not always go together when it comes to the philosophy of science. In the end, reading this book will be a rewarding experience: you will come to understand what scientific explanation is, as well as why philosophy is absolutely necessary for doing science – and for understanding life.

Kostas Kampourakis, Series Editor

Preface

The ancient Greek philosopher Aristotle began his *Metaphysics* by stating "all men by nature desire to know." He was right, but there's much more to the story than that. For one thing, it's not just men who desire to know. All people desire to know. For another thing, we don't want to know in general; we want to know specific things. In particular we want to not only know what has happened, but also *why* it happened, *how* it happened, whether it *will* happen again, whether it can be *made* to happen or not happen, and so on. In short, what we want are explanations. Anyone who has spent even a moderate amount of time around young children knows that two of their first, and favorite, things to ask are "why?" and "how?" This practice isn't something that we abandon as we grow older. In fact, asking and answering "why" and "how" questions lies at the very heart of scientific practice.

How do we answer such questions? By giving explanations. In order to answer why plants need sunlight to survive and how they make use of sunlight to survive, we need to know about photosynthesis, which explains why and how plants make use of sunlight. Natural selection explains why particular traits are present in populations and how particular traits have become predominant in a given population. Molecular polarization explains why and how certain molecules bond to one another. And so on. In each of these cases, we are trying to provide an account of why things are the way they are (and perhaps why they are not some other way).

Explanation is a key aim of science. We seek to explain the world around us because doing so allows us to come to know about the world and to better understand it. The primary aim of this book is to help readers understand how science explains the world. More specifically, the book explores the nature and contours of scientific explanation, how such explanations are evaluated, as well as how they lead to knowledge and understanding. Science helps us understand the world by giving us explanations of the phenomena we encounter. This book is broader in scope than many of the others in the *Understanding Life* series as it is focused on features that one finds in all scientific domains. Nevertheless, as fitting for a book in this series, issues from the life sciences will be given pride of place as the book seeks to facilitate an understanding of how it is that science explains the world.

Acknowledgments

Thanks to Olivia Boult, Samuel Fearnley, Jessica Papworth, Jenny van der Meijden, and all the folks at Cambridge University Press and Integra Software Services for their support throughout the many stages of this project. While writing this book I benefited from insightful feedback on earlier drafts from Stephen Grimm, Kostas Kampourakis, and Michael Strevens. Many thanks! Kostas Kampourakis must be singled out for additional thanks. He not only suggested this project to me, but he also provided helpful guidance and support at every stage along the way. Kostas is a wonderful co-author, colleague, friend, and series editor. Finally, I am grateful to Molly, Kaison, Wallace, and Declan for the love and support that made it possible for me to write this book.

1 Why Explanation Matters in Science

The Primary Aims of Science

While it isn't necessary to do so, it's often good to start a book by saying something that is clearly true. So, let's do that. Science has had (and continues to have) a significant impact upon our lives. This fact is undeniable. Science has revealed to us how different species arise, the causes of our world's changing climate, many of the microphysical particles that constitute all matter, among many other things. Science has made possible technology that has put computing power that was almost unimaginable a few decades ago literally in the palms of our hands. A common smartphone today has more computing power than the computers that NASA used to put astronauts on the Moon in 1969! There are, of course, many additional ways in which science has solved various problems and penetrated previously mysterious phenomena. A natural question to ask at this point is: why discuss this? While we all (or at least the vast majority of us!) appreciate science and what it has accomplished for modern society, there remain – especially among portions of the general public – confusions about science, how it works and what it aims to achieve. The primary goal of this book is to help address some specific confusions about one key aspect of science: how it explains the world.

A first step in getting clearer on *how* science explains the world is to consider *why* science even attempts to explain the world. What exactly does science try to achieve? Or, perhaps putting the question more accurately, what do we (humans) seek to accomplish by employing the methods of science? It is widely accepted that there are three primary aims of scientific activity: prediction, control, and explanation of natural phenomena. Different domains of science emphasize some of these aims more than others. For instance, paleontologists don't spend a lot of time focusing on controlling phenomena, whereas biomedical researchers devote a tremendous amount of effort to controlling infections and diseases. Despite these differences in emphasis, explanation is a common thread linking all these aims. For this reason, it isn't uncommon to hold explanation to be the most important of these three primary aims of science. As the US National Research Council has said, "the goal of science is the construction of theories that can provide explanatory accounts of features of the world."

What makes explanation so important to science? The answer lies in what successful scientific explanations give us: understanding. Very roughly, understanding arises when we grasp how various features of the world depend upon one another. When we come upon the scientific explanation of some phenomenon, our understanding of the world increases. By virtue of this increased understanding, we are often able to better predict and control phenomena. For example, having scientific explanations of why and how something like the 2019 novel coronavirus (SARS-CoV-2, the virus responsible for the COVID-19 pandemic) evolved, helps us to better understand the mechanisms by which this virus reproduces and is transmitted. Of course, once we understand how this virus is transmitted from person to person, we can predict which situations are likely to increase or decrease its spread, as well as when we are apt to see significant rises in the number of infected people. Additionally, this understanding can allow us to put into place guidelines that (if followed!) may help control the spread of the virus. Furthermore, it is understanding of SARS-CoV-2 that has allowed us to produce effective vaccines. Without such understanding, it is difficult to see how we could manage any of these feats.

Considering the role that understanding plays in both prediction and control, it is maybe a bit misleading to characterize science as having three primary aims as we did above. P. W. Bridgman, a Nobel Prize–winning physicist, once said "The act of understanding is at the heart of all scientific activity." Another Nobel Prize laureate, Erwin Schrödinger, claimed that the foundation of the entire modern scientific worldview rests upon the "hypothesis that the display of Nature can be understood." Understanding is central to science, and perhaps it is most accurate to say that the primary *epistemic* (pertaining to knowledge/cognitive success) aim of science is to produce understanding via scientific explanations. Using the understanding gained via scientific explanations to yield accurate predictions and to allow for increased control of phenomena are important secondary aims of science. There are, of course, important caveats and qualifications of this relationship between the goals of science. For instance, science often makes use of models (representations of events/phenomena in the world) in order to explain and predict phenomena. In many cases, however, we might be forced to make choices between models that offer better scientific explanations and models that make more accurate predictions. This trade-off is especially clear when we look at what is called "robustness analysis," which is common in climate science. Robustness analysis involves analyzing a number of incompatible models (i.e., models that make different assumptions about the phenomena being modeled) in order to come up with predictions. In many cases robustness analysis leads to predictions that are considerably more accurate than can be achieved by looking at a single model. However, this often comes at the expense of explanation because we can't really explain what is going on by consulting models that disagree with one another. Hence, at times we seem forced to choose between having better explanations of important phenomena or being able to make more accurate predictions about those phenomena.

For now, we can set aside this and other concerns (we will come back to them later) and consider the general picture of science that emerges when we consider its primary aim(s). Understanding is the central aim of science, and we gain understanding in science by way of scientific explanations. As is often the case, with new insights come new questions. What exactly is understanding? What are scientific explanations, and when are they successful? How do scientific explanations, when they are successful, provide us with understanding? We devote considerable attention to answering these, and many other, questions throughout this book. For the remainder of this chapter, the goal will be to get a firmer grip on the general ideas that will be more fully explored later.

Scientific Explanation

The nature of scientific explanation was a major focus of philosophy of science in the twentieth century, since at least the late 1940s. And, given the difficulties of philosophical analysis of important concepts and the nuances of explanation in science, it is unsurprising that the nature of scientific explanation remains a topic of much debate among philosophers of science even today. Over the course of the history of this discussion there have been many accounts of the nature of scientific explanation and a lot of objections, rebuttals, revisions, and developments of various theories. In fact, there have been too many theories of scientific explanation for us to explore or give much of an overview of even just the most influential accounts here. However, scientists have been using scientific explanations to better understand the world since science began. Furthermore, they have been evaluating scientific explanations to great success – and, there's no reason to think that they'll stop succeeding in this way. Importantly, all of this progress in terms of developing, testing, and evaluating scientific explanations has taken place despite it remaining unsettled which account of the general nature of scientific explanation is correct.

This notwithstanding, a plausible working model of scientific explanation will be helpful to have in hand. This should be ecumenical in the sense that it is at least compatible with the major theories of scientific explanation that have been put forward. Here is such a model: scientific explanation is a matter of tracking dependence relations. The idea here is that a scientific explanation consists of information about how or why one thing depends upon other things. Importantly, this "dependence" view of scientific explanation allows for all sorts of relations – causal relations (when something causes something else), constitution relations (when some things make up something else), mereological relations (relations that exist between the parts of an object), and so on – to count as explanatory. As a result of this, the dependence view of scientific explanation is consistent with all the major views of the nature of scientific explanation that have arisen in the philosophical literature.

Let us consider very briefly one scientific explanation on this model. Consider, for example, cystic fibrosis. People with this disease produce an excess of mucus that can cause passageways in the lungs to clog and obstructions to form in the pancreas. What explains why a particular person has cystic fibrosis? Mutations of the cystic fibrosis transmembrane conductance regulator (*CFTR*) gene. In terms of the dependence view of scientific explanation, we have a (at least partial) scientific explanation of cystic fibrosis. The scientific explanation of a particular person having cystic fibrosis (what is known as the *explanandum* – what is being explained) consists of information about how the person's having this disease is dependent upon other things, in this case the person's having mutated alleles of the *CFTR* gene (this is the *explanans* – what does the explaining, in other words, what provides the information about how or why the *explanandum* depends upon other things).

Here's a very brief recap. Our general picture of scientific explanation is that it consists of information about dependence relations that exist between a particular phenomenon (whether it is a general process or a particular event) and other phenomena. More specifically, we have a scientific explanation of X when we have information about how or why X depends upon some other things, such as Y. In this sense, the explanation can be represented in a question-and-answer format: "Why X? Because of Y," where X is the explanandum and Y the explanans.

Scientific Understanding

We've mentioned that understanding is the key epistemic goal of science. Let's take some time to get a bit clearer about what we mean by understanding in this context. We use the term "understanding" in myriad ways. For instance, we sometimes say things like "I understand that you're angry with me" as a way of expressing a belief that we have while hedging a bit. We're letting the other person know that we think that they are angry with us, but we don't want to fully commit to being correct about this. We might also use "understand" in a way that is synonymous with knowing that something is a fact. "I understand that humans have 23 pairs of chromosomes" is just another way of expressing "I know that humans have 23 pairs of chromosomes." Finally, we might experience an "aha" moment in which a particular scientific explanation feels as if it is giving us insight into the workings of the world. We might be tempted to call this feeling itself understanding. Even if "understanding" is the appropriate term to use for this particular sort of feeling, that is not what we mean when we say that science provides understanding. We're looking for more than just a good feeling because such feelings sometimes are misleading.

The sense of understanding that we're interested in isn't a matter of hedging, merely expressing knowledge of a simple fact, or experiencing a feeling that a scientific explanation is correct. The sense of understanding that matters for our purposes and is the aim of science is a kind of cognitive success. It involves really grasping how the world is – not how we want it to be or what would make us feel good. When we genuinely understand some phenomenon, we are in an epistemic state that may or may not be accompanied by the phenomenal feeling of "aha." Like almost everything else that philosophers study, there are numerous minute differences when it comes to accounts of understanding (and many differences that are not so minute). This shouldn't dissuade us in our discussion though. After all, philosophers have been arguing about the exact nature of knowledge for many years (and they still don't seem to be anywhere near a consensus), but this hasn't hindered our general ability to recognize whether something is known or not.

While there is live debate among philosophers about whether understanding is itself a kind of knowledge, it is generally agreed that understanding is something beyond mere knowledge of facts. To illustrate the plausibility of this idea, consider a student who simply parrots what their biology teacher tells them. This student knows, for example, that the DNA sequences of humans and chimpanzees are extraordinarily similar because the teacher has said this. However, the student may be clueless as to why this is so, or the importance of this similarity within the larger context of biology. The student may have no grasp of how this fact provides support for evolution in general and common descent in particular. What is missing? It seems that whereas the student simply knows this (and perhaps many other) isolated facts, understanding requires seeing how these facts hang together. The person with understanding grasps biology (or any other object of understanding) as an interrelated body of information with many connections between the various facts. This person can appreciate how these various facts depend upon one another. Furthermore, the person who understands biology can use this understanding to explain and sometimes predict particular biological phenomena.

Following philosopher of science Henk de Regt, we can helpfully distinguish between two varieties of understanding. The first is what we've been primarily considering until now: understanding phenomena. We achieve this sort of understanding by coming to know why or how a particular phenomenon occurs. In other words, we come to understand phenomena when we grasp correct scientific explanations of what causes the phenomena, the mechanisms or processes involved in the production of the phenomena, how the phenomena fall under natural laws, and/or how various changes to other things might have led to changes in the phenomena. A classic example of this is the understanding of the variation among the finches of the Galapagos Islands. As Charles Darwin noted, we can explain this phenomenon (variation among the different kinds of finches on these islands) by recognizing that it was the result of adaptation by natural selection. Recognizing the process (natural selection) and the causes (differences in environmental conditions, variation in characteristics such as beak size within populations) that led to the diversification of the finches yields understanding of the phenomena. This sort of understanding is the primary aim of science.

The second sort of understanding that we are concerned with is understanding a theory. One can genuinely understand phenomena only if one understands the relevant scientific theories. What exactly does it mean to understand a scientific theory though? And, what do we even mean by calling something a scientific "theory"? Let's start with the second question. We don't mean by "theory" the sort of thing that is far too often meant by it in public discourse. In such cases "theory" is often used to signify a claim or hypothesis that is still the subject of significant, reasonable doubt. It is exactly this sort of use of "theory" that is operative when critics erroneously charge that evolution is *just a theory*. Instead, when we speak of theories in science, we are talking about well-established domains of science that enjoy strong empirical support and include many widely accepted foundational facts, methods, and laws or principles. When it comes to understanding a theory this consists of being able to use the scientific theory to construct (or at least appreciate) scientific explanations or make predictions about phenomena within a particular domain. For instance, someone who understands evolutionary theory can construct scientific explanations of a number of things such as the variation that one finds among the different species of finches in the Galapagos Islands; or make predictions about where specific fossils could be found, as in the case of *Tiktaalik*, an extinct lobe-finned fish that has many similarities with four-limbed animals.

Possessing understanding of a scientific theory will depend on various factors. Some of those factors have to do with the scientific theory itself – the simplicity of its structure, its facility to be utilized for predictions, and so on. Other factors will depend upon the individual – things like background knowledge and intellectual capacity are key factors when it comes to whether someone is able to come to understand a scientific theory. Of course, understanding is something that comes in degrees. So, one person might have a deeper understanding of a scientific theory (or phenomenon) than another. An easy way to see this is to consider the different levels of understanding that an expert and an informed layperson may have when it comes to particular scientific theories. An expert can generate a new evolutionary explanation of some disease, such as COVID-19, which suggests that the virus SARS-CoV-2 likely evolved naturally rather than being designed in a lab. A layperson can appreciate this explanation when hearing of it, but typically the layperson couldn't come up with this explanation on their own. In such a case, both the expert and the layperson are exhibiting some degree of understanding of theories of viral evolution. However, the expert is exhibiting a significantly higher degree of understanding of the scientific theories in question, and plausibly as a result of this the expert has a deeper understanding of the phenomenon that the scientific theories are being employed to explain – for instance, how the sequences of the genomes of the various viruses can be compared and how such comparisons can form the basis for estimating evolutionary relations.

It is worth briefly pausing to emphasize the importance of distinguishing between understanding phenomena and understanding theories. Two primary reasons this is important to do are that this distinction helps us better appreciate how science achieves understanding and why such understanding really is an *achievement*. As we have discussed, coming to understand phenomena requires exercising one's cognitive faculties, in particular one's understanding of scientific theories, to generate or appreciate scientific explanations of the phenomena in question. For genuine understanding of a phenomenon, it is not enough that one is simply informed of a scientific explanation; one must appreciate how the scientific explanation provides an account of why or how the phenomenon occurs. At the heart of this process lies scientific explanations – generating them, comparing them, or at the very least appreciating them. The central cognitive aim of science cannot be had without scientific explanations.

Key Successes of Scientific Explanation

The history of science is replete with examples of successful scientific explanations. Often, as we discuss in later chapters, these scientific explanations lead to significant new discoveries. In other cases, they provide deep understanding of phenomena that were previously mysterious. And, in many other cases, they help aid in controlling various phenomena (such as infections) and developing new technologies. Arguably a big part of the reason that science advanced so quickly after the scientific revolution began is that key scientific explanations were hit upon. That said, let's take a brief look at two of the incredible successes of scientific explanation (we'll consider others in later chapters).

In the early 1800s, it was discovered that the orbit of Uranus (at that time believed to be the last planet in our solar system) didn't follow the path predicted by Newton's theories, coupled with the assumption that there were no other planets. What was to be made of this? Since the empirical evidence was undeniable, there were only two options. Either give up Newton's theory of universal gravitation or abandon the assumption that there were no other planets beyond Uranus in our solar system. At that time, especially given its tremendous successes, dropping Newton's theory wasn't appealing. Two scientists, John Couch Adams and Urbain Leverrier, working independently of one another, hit upon a better explanation. They postulated that there was a thus far undiscovered planet that was causing the orbit of Uranus to be different than expected. This explanation accounted for the strangeness of Uranus' orbit without abandoning Newton's theory. The great success of this scientific explanation came shortly after it was put forward, when the existence of Neptune was observationally confirmed.

Another great success of explanation in science was the line of reasoning that led Charles Darwin to the theory of natural selection. As he described in *The Origin of the Species*: "It can hardly be supposed that a false theory would explain, in so satisfactory a manner as does the theory of natural selection, the several large classes of facts above specified. It has recently been objected that this is an unsafe method of arguing; but it is a method used in judging of the common events of life and has often been used by the greatest natural philosophers." The explanatory success of evolutionary theory in the life sciences can hardly be overstated. As evolutionary biologist Theodosius Dobzhansky once said, "Nothing in biology makes sense except in the light of evolution." The understanding of various phenomena that has resulted from understanding evolutionary theory is nothing short of astounding.

There are innumerable other instances where scientific explanations have been tremendously successful – the oxygen theory of combustion, the discovery of electrons because of the explanations their existence provides, and many others. Scientific explanations have been tremendously successful as the method of achieving the primary aim of science: understanding.

At this point one might be inclined to wonder: What is the relationship between understanding and truth? Does science ever get to the *absolute* truth? Can we really understand without knowing the whole truth? Does science even aim at discovering the truth? These questions mark a good place to briefly pause to avoid a potential misunderstanding about how truth figures into the discussions in this book. While scientific knowledge and the depth of our understanding is always apt to change over time, this doesn't mean that science doesn't seek truth – it does. The sense in which scientific knowledge changes over time is that we often learn that what we *thought* was knowledge wasn't genuine knowledge. When we speak of the current state of scientific knowledge or our depth of scientific understanding, we are speaking of what we have good evidence to think is actually the truth. However, because that evidence is never sufficient for absolute certainty, our judgment is always open to revision, that is, our judgment of whether something is the absolute truth and whether we fully understand something is tentative.

What's to Come

In this chapter we've properly begun our investigation into how science explains the world. Predominantly, our discussion has so far centered on *why* science explains the world. As we have seen, the reason for this is fairly simple. Science explains the world because it is by way of scientific explanations that the chief cognitive aim of science is achieved. We only come to scientific understanding of the physical world by possessing scientific explanations. In the remainder of the book, we turn our attention to *how* science explains the world. Here's a brief preview of what's to come and the many questions we seek to answer.

In Chapter 2, we discuss the relationship between scientific explanations and ordinary, everyday explanations. How are they similar? How are they different?

We turn our attention to specific kinds of scientific explanations in Chapter 3. How are scientific explanations that describe what actually happens related to scientific explanations concerning what might or could happen in other situations? Can historical explanations be genuinely scientific? If historical explanations can be genuinely scientific, are they, nevertheless, inferior to experimental scientific explanations? Do even the best scientific explanations (whether experimental or historical) explain everything and leave no uncertainties?

Our focus in Chapter 4 is the relationship between scientific explanation and prediction. Are they the same thing, or at least symmetrical? Can good scientific explanations fail to make accurate predictions? Can scientific models that provide accurate predictions fail to offer good scientific explanations?

We explore how we evaluate the quality of scientific explanations in Chapter 5. What sort of features are theoretical virtues? Do these virtues make one scientific explanation better than another? Granting that particular theoretical virtues do make some scientific explanations better than others, are we any good at evaluating which scientific explanations actually have these virtues? What if theoretical virtues are simply things that make models or theories convenient for us to use? In other words, should we think that theoretical virtues are really guides to the truth?

In Chapter 6, we delve more deeply into the relationship between scientific explanations and the understanding that they generate. What is understanding in general? How does one come to have scientific understanding? How is it related to the experience one gets when it *feels* as if one understands?

Next, in Chapter 7, we examine the role that idealizations play in generating scientific explanations. What exactly are idealizations? Why do we use them in science? Can idealizations be a means of generating scientific understanding?

We conclude the main part of our discussion, in Chapter 8, by looking at how scientific explanations generate scientific knowledge. Scientific explanations generate scientific knowledge via inference to the best explanation. But, what exactly is this method of inference? When do we employ inference to the best explanation? How does this inference allow scientific explanations to generate scientific knowledge? Is inference to the best explanation even a legitimate form of reasoning? If a model or theory is to be "real" science, must it not only be the best explanation but also explain everything in a given domain?

In answering the above questions our discussion covers a lot of ground from the nature of scientific explanation to how scientific explanations are used to generate scientific understanding and scientific knowledge to the role of idealizations in producing scientific explanations. We take our time working through these issues so that we can harvest genuine insights. That said, we also avoid getting bogged down in details that are unnecessary for our purposes. By the end of our journey, a clear picture of the role of explanation in science emerges. Along the way we consider, and clear up, various common misunderstandings of these issues (these are summarized at the end of the book). Now let's continue on our path to discovering how science explains the world.

2 The General Nature of Explanation

We Always Look for Explanations

Explanation is central to our lives. We seem to have an innate (or nearly so) drive to explain and seek explanations. When our favorite app is not working, we want to know why, and we want to know how to fix it. When trying to understand why people engage in an odd behavior – refusing to wear masks during the COVID-19 pandemic, say – we want an explanation. What reasons do they have for doing something that seems so clearly misguided? Why are they resistant to expert advice on the issue? Ultimately, we seek explanations to help us understand and navigate the world around us.

Seeking and providing explanations is extremely widespread and begins very early in human development. No one who has spent much time around young children will be surprised to learn that there is a significant amount of psychological evidence suggesting that very early on children exhibit curiosity about just about everything and that they often seek explanations. Like scientists and adult laypeople, children often approach the world by asking "why?" and "how?" One reason for this is that seeking explanations and even explaining to oneself is an important way that people, both children and adults, learn.

Giving and seeking explanations is so common that some, such as philosopher Ted Poston, have argued that "explanation" is a primitive concept. In other words, Poston maintains that explanation is something that humans simply have a grasp of without being able to (or needing to) break down into component concepts. We can just tell whether something explains something else. In support of this primitiveness claim, Poston marshals a variety of evidence from the fact that philosophers have long sought, and failed, to provide an analysis of explanation in terms of more fundamental concepts to the fact that explanation-signifying words like "because" are some of the most commonly used words in various languages (for example, it is among the 100 most common words in English, French, German, and Spanish). Of course, the most telling evidence is that which we have already mentioned – explanation has a central role in our cognitive lives from early childhood onward. As psychologist Tania Lombrozo says, "generating and evaluating explanations is spontaneous, ubiquitous, and fundamental to our sense of understanding."

Regardless of whether we agree with Poston about "explanation" being a primitive concept, it seems that Lombrozo's assertion is undeniable. Explanation is all around us, and it is of vast importance to our lives. But what exactly is it? If Poston is correct, it will be impossible for us to give a successful reductive analysis of explanation. That is of little concern here, though. Reductive analysis of the concept of explanation is not our ultimate goal even if, *pace* Poston, such an analysis is possible. Instead, our goal is to understand how science explains the world, and there is much that we can clarify and come to better understand about explanation without reducing it to other more primitive concepts. That said, let's start looking at the general nature of explanation.

Explaining versus Explanation

The first step to grasping the general nature of explanation is to draw an important distinction between explanation and explaining. The former is what one gives or receives in a successful instance of the latter. For instance, when your friend arrives late to your lunch date, and she tells you about how her meeting went longer than it was supposed to go, and how she then got stuck in construction traffic, and so on, she is explaining to you why she is late. The explanation that she is giving is the information about the circumstances which led to her being late. The general idea is straightforward: An explanation is a set of statements that account for why or how something occurred, whereas explaining is an action that we take in order to provide an explanation.

Although the difference between explaining and explanation is easy to grasp when it is spelled out, it is worth thinking carefully about this distinction as doing so helps us avoid errors. One error is thinking that any time someone engages in an act of explaining, they are thereby offering a genuine explanation. This, of course, is far from the truth. People can try to "explain" phenomena without knowing the first thing about why or how the phenomena occur. The mere fact that someone is purporting to explain something does not entail that what they are saying is actually an explanation of the phenomenon in question.

Another error that can be avoided by clearly distinguishing between explaining and explanation is that of thinking that because someone does a poor job explaining something, there is not a clear or well-understood explanation to be had. How well someone can perform the action of explaining is at least somewhat independent of whether there is a good explanation available – it is even somewhat independent of whether the person knows that good explanation. It is *somewhat* independent because it is true that someone might know a good (even *the* correct) explanation of some particular phenomenon and yet fail to do a good job of explaining because they are unable to convey that explanation to others very well.

Explaining and possessing a good explanation are not completely independent, however. After all, if one does not know a good explanation of a particular phenomenon, then one is not in a position to explain that phenomenon to someone else. When we explain, we are sharing our knowledge of the facts constituting the explanation. We cannot share what we do not have. So, if we do not know how or why a particular phenomenon occurred or has certain features, we cannot transmit that explanation to others.

In sum, in order to be in a position to successfully explain a phenomenon to someone else, we need both to know the explanation of the phenomenon in question and to be able to convey the explanation in an accurate way that is comprehensible to the recipient of our explaining.

Answering "Why" and "How" Questions

We have already seen that an explanation is a set of statements that account for why or how something occurred. We can thus say that an explanation is an answer to "why" and "how" questions. Why was your friend late? How did construction traffic lead to her being late? Why did this particular trait become prevalent in this population? How is it that these particular genes give rise to this particular characteristic? One way to get a handle on what sort of statements are explanatory is to simply think about the word "because." Explanations are statements that provide a "because" answer to "why" or "how" questions. Your friend is late *because* she got held up at work and stuck in traffic. The construction traffic contributed to your friend's being late *because* the additional traffic forced her to drive slower, and she got stuck at more traffic lights. Spiny tails evolved in a particular species of lizard *because* this trait helped earlier generations of lizards survive attacks from predators and go on to reproduce. The spiny tail appeared in some lizards *because* of mutations of particular segments of their DNA which resulted in changes in their development.

Not just any "because" answer will do though. Parents sometimes answer their children, usually after a long stream of questioning, "because I said so," or "because that is just how it is." Such answers, though perhaps accurate, are not very explanatory because they do not provide any understanding. Hence, an explanation is a "because" answer to an appropriate "why" or "how" question, but such an answer is an explanation only when it provides understanding of why or how it is that the phenomenon in question came to be.

At this point, we need to realize that a mere description of a phenomenon is not sufficient for an explanation. In other words, simply describing what happened does not amount to explaining what happened. This is true even though a description might in some sense be taken to answer a "how" question. For instance, it might be accurate to say that a particular species of lizard now has spiny tails by pointing out that in the evolutionary history of the species there is an ancestor with a spine-free tail, then a more recent ancestor with a single or a few spines, and so on. This description may be correct in that the actual history followed this sequence. Nevertheless, this mere description fails to explain *why* or *how* the species developed spiny tails. What is lacking? There is no clarification of what conditions caused this trait to arise or how they caused it. In order to have an explanation we need to know what made a difference to a phenomenon's occurring as it did, rather than some other way, or not at all. We will discuss the idea of difference makers more fully below. For now, it is enough to note that genuine explanations include information about difference makers and how some things depend upon other things.

Before moving on we should touch upon one further distinction that arises when thinking about explanations as answers to "why" and "how" questions: the distinction between explanations that tell us why or how something *actually* came to be and explanations that tell us how something *possibly could* be the case. Although we often focus on the first sort, what we might call "how/why-actually explanations," the second sort of explanations, which are called "how-possibly explanations," also play an important role in science and our everyday lives. A how-possibly explanation does not answer why or how something actually happened, but instead provides information about how it is possible for something to have happened or how something that hasn't happened might happen.

Perhaps the simplest way to get a firm grasp upon how-possibly explanations is to consider a situation like the following: imagine wondering about a past championship. Going into the final two games of the season, Team A was in second place in their division because their record was 11 wins and 1 loss, whereas Team B was in first place with a 12–0 winning record. Assuming that each team had two games left in the season and that the team who finishes with the best record goes to the championship game, it is not difficult to appreciate a how-possibly explanation of Team A's making it to the championship game. How can Team A have possibly made it to the championship game? If Team A won both of their remaining games and Team B lost both of their remaining games, then Team A would have gone to the championship game because they had a record of 13–1, whereas Team B would have a record of 12–2. Other possible explanations of how this could possibly have occurred would be that Team B was disqualified for some reason, such as financial fraud, or that they decided to withdraw from the championship due to having too many injured players. If you do not know what actually happened, you might construct such how-possibly explanations, and then seek confirming or disconfirming evidence.

How-possibly explanations also arise in science both in contexts of theoretical speculation and contexts of great practical concern. Biologists might be concerned with how a species might have possibly evolved to have a particular new trait. In providing information about how the species could come to possess a new trait that it does not have yet, one is giving a how-possibly explanation. Such reflections may or may not have immediate practical implications. Other instances of generating how-possibly explanations are very practical and of critical importance. For instance, many scientists are hard at work generating explanations of the effects of climate change that we are currently experiencing. The understanding of such how-possibly explanations generates suggestions for interventions to help slow or reverse some of the changes that human activity has wrought upon climates around the world. The issue here is that we may not know all the causes of climate change or how they produced the changes we see; however, thinking about how climate change possibly occurred can be instructive for figuring out what we can do about it now.

Must an Explanation Entail What It Explains?

What sort of structure must a collection of explanatory statements exhibit in order to constitute an explanation? According to the twentieth century's most influential account of scientific explanation, explanations are arguments. More specifically, philosopher Carl G. Hempel's *covering law model* of explanation (which though no longer widely accepted is still arguably the most influential picture of the nature of scientific explanation) views explanations as arguments that include premises describing conditions that led to the phenomenon being explained. Among these premises, there is at least one that describes a natural law that connects the conditions described in the other premises to the phenomenon to be explained. We will discuss the covering law model of explanation, as well as why many think it is flawed, because doing so will help us better understand the nature of explanation.

The covering law model admits of two forms of explanations: deductive-nomological (D-N) and inductive-statistical (I-S). These forms of explanations arise from the two basic forms of arguments: deductive and inductive. Successful deductive arguments are such that the truth of their premises guarantees the truth of their conclusions. In other words, a deductive argument for X is an argument that is such that it is *impossible* for all the premises of the argument to be true while X is false. In contrast, inductive arguments never guarantee the truth of their conclusions. Rather, strong inductive arguments are such that when they have all true premises, their conclusions are very likely to be true. Here are simple examples of each type of argument:

Deductive argument:

1. All humans are mortal.
2. Socrates is a human.
3. Therefore, Socrates is mortal.

Inductive argument:

A. Many ravens have been observed in a variety of circumstances.
B. All the observed ravens have been black.
C. Therefore, all ravens are black.

Although both arguments provide strong support for their conclusions (if all their premises are true), only the first provides *conclusive* support for its conclusion. There is no way that 1 and 2 are true and yet 3 is false. However, it is possible, even if very unlikely, that C is false even if both A and B are true.

Deductive-nomological explanations are deductive arguments that entail the truth of a statement describing a particular phenomenon's occurrence. So, in order to have a D-N explanation of something, say a particular piece of lead's melting, the explanation must entail a statement describing the piece of lead's melting. Such an explanation might look like this:

1. Conditions: This piece of lead was heated to above 327.5 °C.
2. Law: Lead melts at 327.5 °C.
3. Therefore, this piece of lead melted.

Of course, not all laws are deductive in this sense. Many natural laws are inherently probabilistic. This is why the covering law model includes I-S explanations. The general idea here is that an I-S explanation of X is an inductive argument that makes it more probable than not (i.e., above 50 percent likely) that a statement describing X's occurrence is true. An I-S explanation might look like this:

A. Conditions: The patient was suffering from a particular illness, Y, and took medication M.

B. Law: 75 percent of patients suffering from Y who take M will recover.
C. Therefore, the patient recovered from Y.

The clear logical structure of the covering law model of explanation and the intuitiveness of the idea that an explanation should make the phenomenon being explained to be expected account for the appeal of this picture of the nature of explanation. Unfortunately, there are good reasons for thinking that it is mistaken.

Let us start with some of the problems for the D-N model. A classic problem involves considering the relationship between barometers and storms. Changes in barometric pressure cause storms, but they also cause changes in the indications of barometers. This means that prior to a storm there will be a change in what a barometer indicates about barometric pressure. On the D-N model, it looks like one can explain a particular storm by noting that the barometer reading changed and that barometer readings always change before a storm. Of course, this is incorrect. The change in the barometer reading itself does not explain the occurrence of a storm. Rather, it is the changes in barometric pressure that explain both the change in the barometer readings and the occurrence of the storm. Change in barometric pressure is a common cause of changing barometer readings and of storms. The D-N model flounders when confronted with such cases of common causes.

Second, the D-N model runs into problems because natural laws are symmetric in important ways. The most famous illustration of the difficulty this generates for the D-N model involves a flagpole. On the D-N model, we can explain the length of the shadow cast by a flagpole by citing the height of the flagpole, the angle of the sun, a bit of geometry, and natural laws about light traveling in a straight line. We can deduce the length of the flagpole's shadow from all the rest. Hence, the D-N model yields that these facts explain the shadow's length. For example, assume that the angle of the sun relative to the flagpole is 30° and that the flagpole itself is 3 meters tall (and the flagpole is standing straight, perpendicular to the ground). In this case, we know that the triangle formed by the flagpole, its shadow, and the sun's rays is a right-angled triangle, and in particular it's a right-angled triangle with interior angles of 30°, 60°, and 90°. Consequently, we know that the height of the flagpole = x^3 and the length of the shadow = x. So, all we have to do to determine the length of the shadow is solve for x. Thus, we can deduce from the available information that the flagpole's shadow is 1.73 meters long. So far, it seems that information about the height of the flagpole, the angle of the sun, a bit of geometry, and natural laws about light traveling in a straight line really does answer why the shadow is a particular length.

However, the D-N model is equally committed to our being able to explain the height of the flagpole based on the angle of the sun, the fact that light travels in a straight line, the same bit of geometry, and the fact that the flagpole's shadow is 1.73 meters long. Hence, the D-N model yields the result that this information provides a good explanation of the flagpole's being 3 meters tall. But this is clearly mistaken. Being able to deduce the flagpole's height in this way provides no explanation of its height. This deduction does not answer why the flagpole is that height or how it came to be that height. Presumably, the actual explanation of the flagpole's height will include information about the building materials that were available, the intentions of those who put the flagpole there, and so on. Thus, explanations cannot simply be deductive arguments that entail that the phenomenon being explained occurs in a particular way – even if those arguments include natural laws as premises.

Now that we have seen why the D-N model is problematic, let us take a look at why the I-S model fails as well. Recall, I-S model explanations do not require that an explanation logically entails that the phenomenon in question occurs. Instead, in order to have a successful I-S explanation one only needs information that renders the phenomenon likely to occur (at least more likely than not). There is a problem here too though – sometimes a particular feature can explain a phenomenon even though it does not make the phenomenon likely to occur. One of the most widely discussed examples of this sort concerns explaining why one person contracted paresis and another person did not. Pointing out that the first person, but not the second, has syphilis explains why the first person developed paresis while the second person did not. The reason for this is that only those with syphilis develop paresis. However, paresis is not common among those with syphilis – the majority of people with syphilis do not develop paresis. Thus, it seems that the first person's having syphilis explains why they contracted paresis when the second person did not, even though this would not count as a good I-S explanation. Similarly, the fact that a particular organism has parents that are both AB heterozygotes (they each possess two different alleles of the gene in question) explains why that organism is an AA homozygote (it possesses two identical alleles of the gene in question). This is true despite the fact that the odds that such parents produce an AA homozygote offspring is only 25 percent, which is far below what is required for a successful I-S explanation. Thus, the I-S model is flawed.

A perhaps deeper problem with the I-S model is that it fails to distinguish actual explanations from mere statistical correlations. A classic example of this problem for the I-S model concerns biological men who take birth control pills. All the men who take birth control pills fail to get pregnant. Given the strong statistical correlation between taking birth control pills and not getting pregnant, the I-S model allows that taking birth control pills is part of the explanation of why these men didn't get pregnant. Obviously though, the fact that they took birth control pills has nothing to do with it! They didn't get pregnant because they are biological men, and men can't get pregnant anyway – whether they take birth control pills or not is irrelevant.

In sum, the covering law model of explanation fails, but its failures are instructive. Simply entailing that a phenomenon occurs is not sufficient for explaining the phenomenon, nor is it necessary. Additionally, an explanation does not have to be such that the truth of the information in the explanation makes the phenomenon more likely than not to occur. In light of the failures of the covering law model, perhaps we should think of explanations more broadly than as providing arguments. Following the demise of the covering law model, there have been many attempts to give a philosophical account of scientific explanation. Fortunately, it is enough for our purposes to recognize two facts. The first is that it is widely agreed that scientific explanation is not different in kind from everyday explanations. This is important because in many cases we can helpfully illustrate truths about scientific explanations by looking at more mundane everyday explanations (this is something we have already been doing throughout this chapter). The second fact is that a promising general approach is to think of explanations as providing information about how the phenomenon being explained depends upon other things.

Explanation as Dependence

In light of the conclusions we reached in the previous section, we are in a position to lay out a helpful working model of explanation to guide our discussion throughout the remainder of this book. As we noted, it is plausible that explanations involve information about how one thing depends upon another; in this sense, explaining is about tracking dependence relations.

One common sort of dependence relation is causal relation. When A causes B, B is dependent upon A. This particular variety of dependence relation is so common that many think that in order for an explanation to be scientific it must provide information about the cause(s) of why something occurred and/ or about the mechanism or causal process that led to the phenomenon in question. Although this view is prevalent, it is not unassailable. Some follow philosopher Elliott Sober in thinking that "equilibrium explanations" are genuinely scientific explanations despite not being causal explanations. The general idea in an equilibrium explanation is that an outcome can be legitimately explained by the fact that a large number of initial states of a system would almost inevitably lead to this outcome. Importantly, such an explanation does not say what particular initial state led to the outcome in question, nor how exactly the outcome was produced in that system. So, such explanations seem to lack causal information.

Sober appeals to Fisher's principle, which offers an explanation of why the sex ratios of most sexually reproducing species is 1:1 (i.e., they have an equal number of males and females), as a prime example of an equilibrium explanation. Basically, Fisher's principle says that if a population of a sexually reproducing species departs from a 1:1 sex ratio there will be a reproductive advantage for pairs that overproduce whichever sex is in the minority. Since sexual reproduction requires both a male and a female of the species, if, for example, the sex ratio is currently so that there are two females for every male, then pairs that overproduce males will have an increased chance of passing on their genes to additional generations. The reason for this is that there are twice as many available mates for male offspring as there are for female offspring. This reproductive advantage will last until a 1:1 sex ratio is reached, after which there will be an equal number of potential mates for offspring of each sex. As a result, neither sex will be more likely to produce more offspring by having multiple available mates. Hence, a large variety of initial sex ratios will eventually lead to a 1:1 ratio because that is the equilibrium state of sexually reproducing populations. Thus, Sober contends that we have a genuine scientific explanation that is, nevertheless, *not* a causal explanation.

It is controversial whether such equilibrium explanations are really noncausal though. Additionally, there is controversy about whether any other sorts of genuine scientific explanations are similarly noncausal. Here we can see an advantage of our working model of explanation. Causal relations are an important variety of dependence relations, but they are not the only sort. Constitutive relations (when some things make up some other thing), relations between parts of an object, familial relations (e.g., the relation that must exist in order for someone to be a person's uncle), and others are all dependence relations. Consequently, the general approach to explanation that we are adopting includes, but is not limited to, causal relations. As a result, we can sidestep the challenging issue of determining whether or not equilibrium explanations and other seemingly noncausal scientific explanations are really causal. After all, equilibrium explanations, whether causal or not, are describing dependence relations. Or, perhaps more carefully, they are describing an *absence* of dependence relations. When the likely outcome of an equilibrium state is independent of the particular initial state of the system in question, grasping this fact can provide a scientific explanation of why a particular equilibrium state is likely to occur in systems of varying initial states.

At this point, it is worth taking a moment to explicitly state what our conception of explanation says. An explanation is a set of statements that provide information about how the phenomenon to be explained depends upon various other things for its occurrence, or for its occurring in the way that it does, or for having the properties that it has. Simply put, explanations answer "why" and "how" questions with information about dependence relations obtaining between the phenomenon being explained and other things (earlier conditions, natural laws, and so on). When we know that whether or not someone has hemophilia A is dependent upon whether their *F8* gene has a particular mutation, we can answer why a particular person has hemophilia A. Additionally, understanding this dependence relation can, along with other information, help us answer how it is that the *F8* gene is passed from a parent to a child and perhaps how it is that mutated *F8* genes lead to the development of hemophilia A. Being in a position to answer such questions by appealing to dependence relations is the same as possessing an explanation of the phenomenon in question.

What If Things Had Been Different?

When we have knowledge of dependence relations, we are also in a position to answer questions concerning how things might have been. Specifically, we are able to answer what philosopher James Woodward has called "what-if-things-had-been-different" questions. For example, when we understand how hemophilia A depends upon a mutation of the *F8* gene we can answer a number of what-if-things-had-been-different questions. One in particular stands out: What if the person's *F8* gene did not have the mutation (i.e., what if the person had different alleles)? The person would not have had hemophilia A. Since hemophilia A is linked to a mutation of a gene located on the X chromosome, and it is inherited in a recessive pattern, we are in a position to answer various other what-if-things-had-been-different questions. For instance, if only one parent has hemophilia A, then we can answer how things would be different depending upon whether or not that parent is the father or mother. Since mothers have two X chromosomes and have hemophilia A only if both X chromosomes have the mutation, when a mother has hemophilia A she will always pass along an X chromosome with a mutated *F8* gene to her offspring. Hence, if a child's mother has hemophilia A, then that child will inherit at least one X chromosome with the mutation. This means that if the child is a boy, he will have hemophilia A because that is the only X chromosome he has; and if the child is a girl, then she will not have hemophilia A because the X chromosome that she inherits from her father in this case will not have the mutation because we are assuming that only one parent has hemophilia A.

What if things were different though? In particular, what if the child's father were the one who has hemophilia A instead of the mother? In that case if the child is a boy, whether he has hemophilia A depends upon whether his mother has one X chromosome with a mutated *F8* gene. Since she does not have hemophilia A herself, it cannot be the case that both of her X chromosomes have the mutation. However, it may be that one of her X chromosomes does. In such a case, there is a 50 percent chance that the boy will have hemophilia. If neither of the mother's X chromosomes have a mutated *F8* gene, then the boy will not have hemophilia A because boys inherit their only X chromosome from their mothers. Similarly, if neither of the mother's X chromosomes have the mutation and the child is a girl, she will not have hemophilia A because she will only have a single X chromosome with the mutated *F8* gene (the one she receives from her father). Again, if the mother does have an X chromosome with the mutation, then the girl will have a 50 percent chance of having hemophilia A because she will definitely inherit an X chromosome with the mutation from her father and there is a 50 percent chance that she will inherit a similarly mutated X chromosome from her mother.

In many cases, knowing answers to what-if-things-had-been-different questions not only demonstrates our understanding of a particular phenomenon, but also allows us to test various potential explanations. We can consider the outcomes that one explanation says should be observed if certain things are different, compared to the outcomes that another explanation says should be observed given those differences. When we manipulate conditions so that they are different in particular ways, we can then check to see which explanation's predictions are borne out. Hence, answering what-if-things-had-been-different questions is key for coming up with testable predictions. Additionally, the more what-if-things-had-been-different questions we know the answers to, the better position we are in to predict, and perhaps control, phenomena.

For instance, recognizing that SARS-CoV-2 makes use of various enzymes in replicating RNA allows one to answer what-if-things-had-been-different questions such as: What if one of the key enzymes the virus needs in order to replicate RNA is interfered with in a particular way? Presumably, the answer is that this can make it harder for SARS-CoV-2 to reproduce in a person's system. The drug remdesivir interferes with one of the key enzymes that SARS-CoV-2 uses in reproducing, so this suggests that giving those infected with SARS-CoV-2 remdesivir may inhibit the virus' ability to multiply. Of course, an alternative hypothesis is that remdesivir does not have this effect. Randomized clinical trials allow us to test the hypothesis that remdesivir has this effect by manipulating whether or not someone infected with SARS-CoV-2 is given remdesivir, and checking to see if the virus multiplies at a similar rate in those given remdesivir and those without any treatment. Answering what-if-things-had-been-different questions is a critical element in achieving some of science's most important aims. Learning how things depend upon one another is how we come to know the answers to what-if-things-had-been-different questions.

Scientific Explanations versus Everyday Explanations

Albert Einstein once said, "The whole of science is nothing more than a refinement of everyday thinking." Whether or not Einstein is correct about the "whole of science," his point squarely hits the mark when it comes to explanations. Seeking and sharing information about dependence relations is not only something that we do when conducting scientific research or providing instruction about science, but is also something that we do in everyday life.

Although most philosophical theories of explanation are commonly referred to as theories of "scientific explanation," explanations in everyday life are roughly continuous with scientific explanations. Granted, the explanations that one encounters in professional science journals and scientific discourse tend to be much more precise and rigorous than the explanations we find in ordinary nonscientific contexts. Nevertheless, the differences are a matter of degree, not a matter of kind. In everyday life, just as in science, explanations offer information about dependence relations. Whether we are in a lab or in a coffee shop, explaining something involves giving information about how that thing depends upon other things.

The close connection between everyday explanations and scientific explanations should not really surprise us. After all, the practice of giving and evaluating explanations is ubiquitous in our everyday lives. We explain why things happen, we seek explanations of others' behaviors, we offer explanations of our own behavior, and so on. Not only this, but as we will see more clearly in later chapters, the reasoning and methods of scientists are quite similar to our everyday reasoning and methods – particularly when it comes to explanatory reasoning. Obviously, scientific methods tend to require more exactitude, and they sometimes lead to rather surprising conclusions from the standpoint of commonsense. Nonetheless, our everyday reasoning shares many of the same commitments as scientific reasoning: that there is objectivity, truth, and rationality. As we discuss later, a number of studies suggest that some of the methods which scientists use to identify dependence relations are plausibly innate methods that we use to understand the world around us.

Whether we are conducting science or engaged in our daily activities, we are giving and evaluating explanations in a similar way. Explanations in scientific contexts and ordinary life are so similar that James Woodward has noted that

> the tendency in much of the recent philosophical literature has been to assume that there is a substantial continuity between the sorts of explanations found in science and at least some forms of explanation found in more ordinary non-scientific contexts, with the latter embodying in a more or less inchoate way features that are present in a more detailed, precise, rigorous etc. form in the former.

For example, when a mechanic explains why a particular car won't start, they do so by citing dependence relations. The car won't start because the battery is dead. This explains why the car won't start despite having a full tank of gas and no other obvious malfunctions. Similarly, a scientist explains why a person has sickle cell disease by noting that the person's *HBB* gene has a particular mutation. In both cases "why" questions are being answered by information about dependence relations. Philosophers of science have long assumed that a successful theory of scientific explanation will be one that makes clear the common features of scientific and ordinary explanations. Thus, it is important to refrain from thinking of scientific explanation as radically different from everyday explanations. Explanation, whether it is scientific or mundane, consists of information about dependence relations.

3 Specific Kinds of Scientific Explanations

Biological Explanations

Although it is helpful to appreciate the general nature of explanations, we might reasonably want more than this. As this book is part of the *Understanding Life* series, we may expect to delve into details about kinds of explanations that are specific to the life sciences. Biologists, however, do not have a unified set of aims or explanatory practices. Hence, when we dig deeper than our general account of explanation, we are apt to find a variety of different kinds of explanations at play in the life sciences. This is the case more broadly in science as well because there is no single scientific method. Some domains of science are dominated by controlled experiments. In other areas, such as paleontology, controlled experiments are rare not only because they are difficult to conduct, but also because they are not of crucial importance. Scientists working in some fields make heavy use of advanced mathematics, whereas important discoveries in other fields, such as cell biology, did not rely on such tools. That said, despite this variety, there are several insights to be had by examining some specific sorts of explanations that one finds in the life sciences. Clarifying and reflecting upon the features of these kinds of explanations can help us better appreciate some of the key sorts of scientific explanations that one does encounter in biology.

Actual-Sequence Explanations versus Robust-Process Explanations

The first distinction to draw between kinds of scientific explanations common to biology concerns the sort of information a scientific explanation might provide about dependence relations. Following philosopher Kim Sterelny, we might call scientific explanations that provide information of the particular features leading to the occurrence of an event or a particular phenomenon "actual-sequence explanations." These are genuine scientific explanations because they provide information about dependence relations, which allows us to answer relevant "why" and "how" questions. However, such explanations are limited in the sense that they often do not contain much information about other possible scenarios. That is to say, an actual-sequence explanation of a particular phenomenon will convey what caused that phenomenon, but it will not provide much information about other ways that the phenomenon might have occurred. Scientific explanations of this latter sort are "robust-process explanations." Like actual-sequence explanations, robust-process explanations provide information about dependence relations. However, unlike actual-sequence explanations, robust-process explanations tend to focus on dependence relations that account for a phenomenon not just in the way it actually occurred, but in other possible situations as well. Let us consider an example to help make these two kinds of scientific explanations clearer.

Imagine that we are seeking an explanation of the biggest blowout in US college football history – the 1916 game in which Georgia Tech defeated Cumberland College 222 – 0. One explanation of Georgia Tech's winning this game would simply point out what happened during each play of the game. This actual-sequence explanation would provide an explanation of the outcome of the game. A different sort of explanation would point to the fact that Georgia Tech had a championship team coached by John Heisman (the person the prestigious Heisman Trophy is named after), whereas Cumberland College had discontinued its football program that year but was contractually obligated to play against Georgia Tech, which led to Cumberland College's roster consisting of students who were not trained players of the sport. This second explanation is a robust-process explanation. It provides information that shows that Georgia Tech's victory would have happened even if the particular plays they made in the game had been very different. For instance, the robust-process explanation provides information that allows one to see that even if the Georgia Tech offense only ran with the ball without the quarterback even attempting to pass the ball to receivers, they still would have won the game.

Although both actual-sequence explanations and robust-process explanations are genuine scientific explanations, they provide very different kinds of information. In some instances, we may want actual-sequence explanations, such as when we are trying to determine the method for reproducing a chemical agent. But in other situations we may be looking for more robust information about different ways things could be, such as when we are seeking an equilibrium explanation for sex ratios in a population, for example. Hence, whether we are looking for actual-sequence explanations or robust-process explanations will depend upon our aims.

Experimental Explanations versus Historical Explanations

There is another important distinction between two kinds of scientific explanations that are often, erroneously, thought to differ in terms of their inherent quality: experimental explanations and historical explanations. In simple terms, experimental explanations are scientific explanations that one finds in experimental settings, and historical explanations are scientific explanations that one finds in settings where the phenomenon in question is either long past or the result of a cause (or causes) that is long past.

In experimental settings, scientists make predictions based on a particular hypothesis and then use laboratory equipment to isolate confounding variables to see whether the hypothesis is confirmed. This is common in biology, chemistry, physics, medicine, and so on. For instance, a new cancer treatment might be tested on mice in a lab setting to see whether it has the desired effect and whether it causes serious side-effects before the treatment is tested on humans. Typically, the sort of explanation that is generated in an experimental setting is one that identifies some repeatable generalization. That is to say, experimental explanations often provide information about what we should expect to happen any time we intervene in particular ways upon a particular substance or phenomenon. They will often be of roughly the following form when dealing with substance X: manipulating X will result in Y. Let's develop the cancer treatment example a bit further. Assume that we are able to explain the death of cancer cells in a group of mice who received treatment with a particular chemotherapy drug by noting that apoptosis (essentially cell suicide) occurred significantly more often in their cancer cells than it did in the cancer cells of a control group that did not receive the drug. In such a case, the experimental explanation provided by studying the effects of the chemotherapy drug on the cancer cells of the mice (e.g., that the chemicals in the drug interfered with various cell processes) could be used to generate predictions. Notably, such an explanation would allow us to predict that the drug being studied would activate apoptosis in the cancer cells of other mice similarly treated with it. Once additional experiments have been conducted, we may have reason to think that the drug will work in other mammals as well.

Historical explanations are different. Take a paradigm historical explanation: the Alvarez meteorite-impact hypothesis which explains the K–Pg (Cretaceous–Paleogene) extinction. This hypothesis explains the extinction of nonavian dinosaurs at the end of the K–Pg period as well as the high levels of iridium in the Earth's layer dated to that time. Or, think about the Pangaea hypothesis, which claims that the continents of the world were once all part of the same super-continent. This hypothesis explains otherwise surprising findings in rocks discovered in various places around the world that are currently located far from one another. Notice, in neither of these cases are we able to experimentally control a particular variable or reproduce the phenomenon in question. As a result, they appear to be significantly different from experimental explanations (though we will soon see that this issue is not as cut and dried as it may appear at first). For now, it is enough to recognize that although we often take experimental explanations as exemplars of scientific explanations, there have been (and likely will continue to be) many very prominent historical explanations in science.

Are Historical Explanations Scientific?

Despite the fact that there are a number of instances of widely accepted historical scientific explanations, one might wonder whether they are really scientific. After all, since we cannot reproduce the phenomena that these explanations appeal to, one might question what makes them different from mere speculation. As editor of *Nature*, Henry Gee once said of historical explanations, "they can never be tested by experiment, and so they are unscientific . . . No science can ever be historical."

It is important to recognize that there is an underlying assumption at play in the general challenge to the scientific status of historical explanations. This assumption is made quite explicitly in Gee's remarks – any hypothesis that cannot be experimentally tested cannot be scientific. Why should we accept this assumption though? It is notoriously difficult to give a precise account of what separates science from nonscience. In fact, this so-called "demarcation problem" is so vexing that many philosophers of science think that it simply cannot be solved. In light of this, it is at least not obvious that we should accept that experimental testability is a necessary feature of the scientific. Let us set aside this concern though, and accept that testability is a requirement for being genuine science. Does accepting this commit us to Gee's position? The answer seems to clearly be "no." For one thing, we need to know why it is so important that the hypothesis be *experimentally* testable. After all, historical explanations make testable predictions. For instance, evolutionary theory predicts that there are transitional species, which have some traits common to ancestral species and some traits common to descendant species. Hence, evolutionary theory predicted that there was a transitional species lying between fish and four-legged vertebrates (such as amphibians and reptiles) before evidence of any such species was found. This prediction was confirmed in 2004 when paleontologists Edward Daeschler, Farish Alston Jenkins, and Neil Shubin discovered fossil remains of *Tiktaalik* (a fish with a number of transitional features, such as bones that would have allowed it to use its fins to prop itself up in shallow water and to use them in a way that is similar to how most four-legged vertebrates use their limbs). It is far from clear why we should think that this kind of testability does not satisfy the requirements for being scientific. For another thing, what is testable (experimentally or otherwise) can change over time. It may be that limits of current technology make it so that a particular hypothesis cannot be tested. However, future advances can arise so that what was once untestable later becomes testable. Finally, there is the simple fact that if we were to accept Gee's very restrictive view of science as only that which is experimentally testable, we would have to deny large swathes of widely recognized sciences as being genuinely scientific. We would have to say that much of astronomy, astrophysics, geology, and paleontology is unscientific.

Aren't Experimental Explanations Superior to Historical Explanations?

When it comes to experimental explanations, we can run various experiments to gather more evidence that might support the explanation. Since we cannot experimentally test historical explanations, should we think that our evidential situation when it comes to historical explanations must be weaker than our situation with respect to experimental explanations? No, we shouldn't.

The ability to test a hypothesis's experimental explanations does not automatically amount to a reason to favor those explanations over historical explanations. Just because a particular experimental explanation is such that we *can* in principle test it in the lab does not mean that the explanation *has been* tested. And, even more importantly, the fact that an experimental explanation can be tested in a lab setting does not entail that the experimental results will in fact support the truth of the hypothesis. Hence, experimental explanations, while testable in lab settings, may have not been tested (in which case the testability does not amount to any more evidence for the explanation) or may have not been borne out by experiments (in which case there is evidence against the explanation). If an explanation is experimentally tested and confirmed, we have evidence that supports the explanation. This is what matters – evidence or the lack thereof – when it comes to favoring a hypothesis and the explanations it generates.

As we see in later chapters, a lot of evidential support for hypotheses comes from their being able to explain various data. In fact, it is not an exaggeration to say that inferences to the best explanation (roughly, inferences where we infer that a particular hypothesis is true because it best explains the data) is at the very core of science. Some historical explanations, such as those provided by evolution, are among the absolute best examples of the power of inference to the best explanation. In fact, Darwin's argument in support of natural selection is often taken to be a prime exemplar of inference to the best explanation. As Darwin's argument makes clear, historical explanations can be very strongly supported by the fact that they provide high-quality explanations that are superior to their rivals. Of course, explanations are not only supported by the data that they explain, as evidence for an explanation can also come by way of support for related hypotheses. This is especially evident when we are thinking of something like evolution, which includes more than one hypothesis. For example, simplifying things a bit, we can understand evolution as the combination of common descent (all living organisms descended from a single common ancestor) and natural selection (some individuals tend to survive and produce more offspring than others). Evidence in support of natural selection clearly supports evolution because natural selection is part of evolution. However, it may also support common descent. How so? The more evidence that we have for natural selection, the better grounds we have for thinking that two species might have diverged from their common ancestors through such a natural process, which in turn supports common descent. To put the general point a bit more abstractly: if a theory entails two hypotheses, then evidence in support of either hypothesis provides evidence in support of the theory, which in turn makes the other hypothesis also more probable. The more probable that evidence makes the theory, the more probable anything entailed by the theory is.

It may be that in some ways it can be easier to gather relevant evidence for (or against) an experimental explanation than it is to gather evidence for a historical explanation, but this does not mean that experimental explanations are inherently superior to historical explanations. At the end of the day, what matters is the evidence that we have in support of a hypothesis, not whether the explanations it yields are experimental or historical.

How Can We Really Know That a Historical Explanation Is Correct?

Even if one agrees that experimental explanations are not inherently superior to historical explanations, one might still worry that there is an important way in which the latter are problematic. When it comes to experimental explanations, one might think that we can know whether they are correct or not by running experiments and testing them. If the experiment turns out differently than the explanation predicts, then we can know that the experimental explanation is false. However, there does not seem to be a corresponding way to test and confirm a historical explanation. As a result, one might worry that we cannot really know whether a historical explanation is correct.

This sort of worry, however, is misguided for a number of reasons. First, it is far from clear that experimental explanations can be directly falsified by a single experiment. The Duhem–Quine thesis (named after the physicist Pierre Duhem and the philosopher W. V. O. Quine) asserts that it is impossible to test a hypothesis in isolation. In other words, any time we test a particular hypothesis, we always do so in conjunction with various auxiliary hypotheses that allow us to formulate predictions. When an experiment yields results that run contrary to our predictions, we might abandon the hypothesis that generates the experimental explanation we are testing, but we are not forced to do so. After all, we could abandon an auxiliary hypothesis instead.

For instance, imagine a situation where we hypothesize that a particular substance is acidic. As a result of this hypothesis, we come up with an experimental prediction – the pH strip that we are about to place in the substance will turn some color on the red end of the spectrum. When we remove the pH strip and it's green, we might conclude that the substance isn't acidic. However, we might instead conclude that the experiment shows us that the pH strip is past its shelf life. Relatedly, sometimes an experimental result is taken to be evidence against a particular hypothesis, but other times that result is simply taken to be an anomaly. Thus, it seems that the picture of science as proceeding by simply discarding hypotheses in the face of contrary experimental results is overly simplistic. Recognition of this is what led philosopher and historian Thomas Kuhn to make his famous case that science proceeds by way of dramatic paradigm shifts rather than the simple rejection of hypotheses after isolated experiments.

One might nevertheless think that with an experimental explanation we can continue to run experiments accumulating more and more data in support of the explanation or against it. And, even if we cannot run a single experiment that will tell us whether or not the experimental explanation is true or false, we can continue to run different experiments that rely upon different auxiliary hypotheses. Consequently, one might argue that with respect to experimental explanations we can continually rule out various auxiliary hypotheses when we encounter results that run contrary to the experimental explanation we are testing. And so, eventually, we can come to know whether the experimental explanation provided by a particular hypothesis is correct or not. But we cannot seem to do this with historical explanations in general. Perhaps this means that we can come to know that an experimental explanation is correct, but we cannot come to know that a historical explanation is correct?

In answering this question, we must not forget that sometimes we use the same methods when engaged in historical investigations as we do when conducting laboratory research. We are often inferring causes on the basis of effects and known mechanisms, for instance. Additionally, we often rely upon historical information, such as data collected from fossils or past observations, in order to inform our experiments and predictions. So, the distinction between experimental and historical explanations is not as clear-cut as it might at first appear. Nevertheless, the possibility of manipulating and controlling variables is a key feature of experimental science that isn't present in historical science. In light of this, one might wonder how we can know that explanations where we cannot utilize repeatable experiments to support them are correct. The answer to this depends upon what it takes to know that an explanation is correct.

Here it will be helpful to make a brief foray into the realm of epistemology. It is widely accepted by epistemologists (philosophers who study the nature of knowledge and related phenomena) that in order to know something, it must be true, and one must believe it is true. Knowledge requires more than this though. Imagine that someone happens to guess the winning numbers in a large lottery. Assume that the person's guess happens to be true and they really form the belief that these specific numbers are the winning ones (before learning the results of the lottery). Does this person know, before the lottery results are revealed, that the winning numbers are the ones that they guessed? Surely not. After all, this person has no good reason to believe that these are the winning numbers. They are simply guessing. And, on top of that, the person is fully aware that they have no reasons for thinking that this combination of numbers rather than some other is the winning combination. Therefore, not only does this person fail to know that these are the winning lottery numbers, but also the person's believing that these numbers are the winning ones is irrational. Given that the person has no relevant evidence concerning which numbers will win the lottery, the rational thing is to not believe that they are the winning numbers. Of course, it would not be rational for this person to believe that these particular numbers are *not* the winning ones either. The person does not have evidence for thinking this either. The person simply has no good evidence to believe of any particular set of numbers that they will win or that they will lose. At best the size of the lottery makes it rational for the person to believe of any given combination of numbers that it will probably lose because this is the most likely outcome (this is assuming that only one combination of numbers out of a very large set of possible combinations wins the lottery). In such a case, the person cannot know whether the numbers they guessed are the winning numbers. This is true even if the person thinks that these are the winning numbers and purely by luck happens to be correct.

As we have seen, simply having a true belief is not sufficient for knowledge. Something more is needed, but what? The ancient Greek philosopher Plato claimed that what is needed besides true belief is some account of the belief's truth which *tethers* one's belief to the truth. The most plausible way of understanding Plato's idea of a tether is in terms of evidence. More specifically, what makes the difference between a *mere* true belief like the lucky lottery guess and a true belief that amounts to *knowledge* is that the latter, but not the former, is supported by sufficiently strong evidence. This is what is sometimes called the traditional account of knowledge: one has knowledge of some proposition when the proposition in question is true and one believes it on the basis of sufficiently strong evidence.

Returning to our question, we can know that a historical explanation is correct when it is in fact correct *and* we have sufficient evidence for believing that it is correct. This still leaves us facing a big question: How much evidence does it take in order to have "sufficient evidence" for knowing that an explanation is correct? One answer is that we must have evidence that is so strong that it would be impossible to have that evidence if the explanation were incorrect. Although such a guarantee is desirable and would be great to have, it is much too strong of a requirement for knowledge. It seems that insisting on such powerful evidence for knowledge leads straightforwardly to an extreme form of skepticism. After all, it is widely accepted that at least at times our senses can lead us astray. Consider a couple of common perceptual illusions. First, think about the phenomenon of experiencing an inferior mirage. This is the common mirage that occurs when it looks like there is water on a hot road when in fact there is not. Another famous illusion is the Müller–Lyer lines. In this illusion, lines of the same length appear to have different lengths. The fact that such illusions exist and are common means that our vision fails to provide us with evidence that guarantees the truth of what we think we see. (Our other senses are susceptible to illusions as well.) Since our vision definitely leads us astray in some cases, it seems possible that in any given case it *could* be leading us astray without our knowing it. Granted the odds that we are the victims of an illusion at any given time are low. Nevertheless, *if* knowledge requires that there is *no chance* that we have the evidence we do while being misled, then our vision (and other senses) never give us knowledge. Obviously, such extreme skepticism is implausible. So, if requiring evidence that guarantees the truth of an explanation for knowledge that the explanation is correct leads to this sort of skepticism, we have good reason to think that evidence that absolutely guarantees truth is not a genuine requirement for knowledge. We need sufficiently strong evidence in order to have knowledge, but the evidence does not have to guarantee the truth of what is known in this strong sense of "guarantee."

So, knowledge cannot be had in the absence of evidence (lucky guesses do not count as knowledge), but we have also seen that knowledge does not require evidence that is so strong that it makes it impossible for the explanation in question to be wrong (requiring such evidence for knowledge lands us in a thoroughgoing skepticism). It appears that knowledge requires something between no evidence whatsoever and maximally strong evidence. There is no obviously correct place to draw the line when it comes to having sufficiently strong evidence for knowledge – in fact, this issue is still widely debated among epistemologists. However, bearing this controversy in mind, we might tentatively adopt what is sometimes referred to as the "criminal standard" of evidence. Drawing from the standard of evidence required for conviction in criminal litigation in the United States, this standard holds that knowledge requires evidence that makes the truth of what is known beyond a reasonable doubt. So, for example, in a normal case of perception, when you apparently see a tree, your evidence (the visual experience of seeing a tree) makes it beyond a reasonable doubt for you that there is a tree. There is a very slight chance that your vision is malfunctioning, or that you are seeing a very cleverly crafted fake tree. Nevertheless, without any evidence for thinking that you are misled by your vision or that there are likely to be fake trees around, your visual experience makes the truth of there being a tree beyond a reasonable doubt for you. Admittedly, "beyond a reasonable doubt" is not as precise as we might want. Nonetheless, it is clear enough for our purposes, and we can take it as our standard for knowledge.

Putting all of this together, we are in a position to answer the question of this section. We can know that a historical explanation is correct by possessing evidence that makes the truth of that explanation beyond a reasonable doubt. This is a level of evidence that we often have in support of historical explanations. For example, the scientific community has reached consensus concerning the truth of historical explanations such as the Big Bang theory and the extinction of nonavian dinosaurs. These historical explanations are supported by a wealth of evidence. And, as we will see in later chapters, such theories possess numerous theoretical virtues, such as beauty, explanatory power, simplicity, and so on, which contribute to their being beyond a reasonable doubt.

But, Can We Be Certain?

One might admit that historical explanations and experimental explanations can be beyond a reasonable doubt, but still worry that we cannot be *certain* that they are true. Whether this is correct depends upon what we mean by "certain." There are at least two important conceptions of certainty that should be distinguished. The first, what we might call "psychological certainty," concerns how confident we are that something is true. Unfortunately, people can be psychologically certain about things for which they have very little evidence. We all have blindspots and biases that at least sometimes lead us to err. One can be psychologically certain of things that are not well supported by the evidence or psychologically uncertain of things that are well supported by the evidence. Since psychological certainty comes apart from evidence possessed, it is not worrisome if explanations are not certain in this sense. In fact, one can argue that being psychologically uncertain is a motivation for further inquiry.

In light of this, if there is a worry here it must lie with the other sort of certainty: "epistemic certainty." Epistemic certainty is not a psychological matter. Rather, it concerns the strength of one's evidence. In order to have epistemic certainty that an explanation is correct, one must have evidence that so strongly supports the correctness of that explanation that it is *impossible* for that explanation to be incorrect given that one has that evidence. When it comes to epistemic certainty, it is true that no explanation is certain. Results always lie within a particular margin of error. There are always questions that remain to be answered. In short, science is full of uncertainties.

Unfortunately, we sometimes mistakenly think that genuine science cannot be uncertain, or that a theory must answer all questions with no uncertainty or else it is "just a theory." A prominent example of this error occurs when critics of evolutionary theory claim that it is "just a theory." When critics assert this, they are expressing skepticism as to whether evolution is true, or can be known to be true. There are perhaps many reasons for why critics charge that evolution is "just a theory," from political to religious to simple misunderstanding concerning what evolutionary theory actually says. One reason that is often given, however, is that evolution does not have an answer for how life began or that there are "gaps" in the evolutionary picture (we do not have fossils of all the species between humans and our common ancestor with apes, for example). Broadly speaking, the concern here is that there are a number of things about which biologists are uncertain when it comes to evolution, and it is likely that there will always be uncertainties, so critics claim that evolution is "just a theory" rather than something that we know to be true.

In a sense the critic is correct – we cannot be epistemically certain that evolution is true. But this should not bother us in the least. No scientific theory, law, hypothesis, point of data – nothing at all in science – is epistemically certain! This standard is so high that we fail to meet it for things that we obviously know. For instance, you are not epistemically certain that you are reading this book. After all, there is a very slight possibility that you are having a vivid dream of reading this book. Obviously, this possibility is preposterous and extremely unlikely. Nonetheless, it is a possibility. And, since it is possible (in the broadest sense of the term) that you have the evidence you do and yet that you are not reading this book, it is not epistemically certain that you are reading this book. Should you panic? Of course not. You know you are reading this book. The thing to conclude here is simply that epistemic certainty is much too high of a standard for scientific knowledge. Hence, it is no threat at all to evolutionary theory (or any scientific explanation/theory) to say that it is not epistemically certain. What matters is whether we have good evidence in support of evolution – evidence that makes its truth beyond a reasonable doubt – and we do have such evidence. There is a wealth of evidence from successful predictions that evolutionary theory has produced. What is more, evolutionary theory is tremendously well supported by its ability to explain the relevant data. As we will see in later chapters, explaining the world in this way is how science gives us knowledge.

4 Explanation and Prediction

Aren't Explanation and Prediction the Same?

In this chapter, we're looking at the relation between scientific explanations and predictions. It is tempting to think that the only difference between explanations and predictions is that one looks back and tells us how or why things happened as they did, and the other looks forward and tells us how or why certain things will (or are likely to) happen. This thought can seem particularly plausible when we consider that in many cases a good scientific hypothesis will both explain phenomena and allow us to make accurate predictions. Despite its initial plausibility, the idea that explanation and prediction are symmetrical is mistaken. The way to see this is to take a look at a particular theory of scientific explanation that entails this relationship between explanation and prediction. The particular theory of scientific explanation in question, the *covering law model,* which we discussed in Chapter 2, is false. One of the reasons that this theory of explanation fails helps illustrate the fact that explanation and prediction are not symmetrical.

Let's first very briefly recall what the *covering law model* says is required for a scientific explanation. On the *covering law model*, a scientific explanation is a deductively valid argument in which statements describing general scientific laws and particular facts (conditions) are premises in the argument, and these premises logically entail a statement which says that a particular phenomenon occurs or has the particular features that it does. In order to see how the *covering law model* leads to scientific explanation and prediction being two sides of the same coin, let's take another look at the two examples of *covering law model* explanations we discussed in Chapter 2. In the first example a particular piece of lead has melted and the explanation of this phenomenon is the following:

1. Conditions: This piece of lead was heated to above 327.5 °C.
2. Law: Lead melts at 327.5 °C.
3. Therefore, this piece of lead melted.

The second example was statistical, but it had the same general structure. A patient has recovered from a particular illness, Y, which is explained in the following way:

A. Conditions: The patient was suffering from a particular illness, Y, and took medication M.
B. Law: 75 percent of patients who suffer from Y and take M recover.
C. Therefore, the patient recovered from Y.

Since *covering law model* explanations are deductively valid arguments, it is easy to see how they could generate predictions of what will happen in addition to explanations of what has happened. For example, given the *covering law model* explanation that we have for why a particular piece of lead melted, it is easy to predict what will happen to other pieces of lead. We can readily predict that if we have another piece of lead and heat it to 327.5 °C or higher, it will melt. The reason is that if it is a law that lead melts at 327.5 °C, then it is true that a piece of lead heated to that temperature or higher will melt. Similarly, if it is true that 75 percent of patients suffering from Y who take M recover, then we can predict that if a particular patient suffering from Y is given M, they have a 75 percent chance of recovering. Explanation and prediction are really the same thing as far as the *covering law model* is concerned. The only difference is that with a scientific explanation we already know the phenomenon being explained has occurred, whereas with a scientific prediction we don't.

Despite the nice and neat picture of the relationship between scientific explanations and scientific predictions that the *covering law model* provides, as we saw in Chapter 2 it has significant problems. One of the key problems with this view of scientific explanation reveals that in general scientific explanations and predictions are importantly different. Recall the flagpole example, but with a slight variation. Imagine this time that we have just measured the length of the shadow cast by the flagpole, but we haven't yet looked closely at the flagpole or measured its height. In this case, we can predict the height of the flagpole, and if our measurements of the flagpole's shadow and the angle of the sun relative to the flagpole are accurate, our prediction of the flagpole's height will be spot on. Nevertheless, despite our extremely accurate prediction of the height of the flagpole, it seems that we have no explanation for why the flagpole is the height that it is or how it came to be that height. The facts that allow us to make our prediction (the length of the flagpole's shadow, the angle of the sun, that light travels in a straight line, and some geometry) have nothing to do with the flagpole's being a particular height. The height of the flagpole is instead explained by things like the intentions of those who put the flagpole where it is, what sorts of material the manufacturers of the flagpole used to construct it, and so on.

In the flagpole example, we have prediction without explanation. Consequently, it turns out that we might be able to predict without explaining. Although explanation and prediction are different, as we will see they are closely related in important ways. Before looking at their close connection, let's first clear up a common misunderstanding about prediction and then see why it is that prediction tends to come apart from explanation.

Is Prediction Only About the Future?

In our ordinary usage of the term "prediction" or "predict" we tend to automatically assume that what is predicted must lie in the future. This is what makes the claim, attributed to baseball great Yogi Berra, "it's tough to make predictions, especially about the future" so funny. We naturally assume that if something is a prediction, then it is about the future. Despite initial appearances, this isn't actually so.

Whether a hypothesis predicts something isn't really about whether the data are yet to be gathered or not. Hence, it is possible that we already have a set of data when a particular hypothesis is generated, and yet that hypothesis genuinely predicts that data. What matters is whether the hypothesis was *designed* to accommodate the data in question or not. If the hypothesis is generated without making use of the data in question, then the fact that the hypothesis entails or makes probable that data is sufficient for that hypothesis to predict that data. For example, the peculiarities of Mercury's orbit around the sun and the fact that its orbit deviated from what Newton's theories said it should be were known for hundreds of years before Einstein formulated the theory of general relativity. Nonetheless, one of the major pieces of evidence in support of general relativity was that it *predicted* the observed peculiarities in Mercury's orbit. Even though the data concerning Mercury's orbit were known well in advance of Einstein's development of general relativity, his theory predicted these data because Einstein didn't rely upon data about Mercury's orbit to construct the theory. In other words, general relativity wasn't designed to entail or make probable any particular facts about Mercury's orbit; it simply turned out that it did make probable otherwise puzzling facts about the way Mercury travels around the sun. Recognizing that the mere timing of when a hypothesis is formulated doesn't fix whether it predicts some data not only helps us to better understand the nature of prediction itself, but also to see why it might be easy to conflate such predictions with explanations.

Good Predictions with Bad Explanations; and Good Explanations with Bad Predictions

We've seen that there are times when we might be able to predict something without having an explanation for it. This is something that happens in legitimate science as well as in our simplistic example of the flagpole. For instance, many climate change models are good at predicting what changes to the global climate we will observe in the future. There is significant agreement concerning the fact that humans are a major cause of climate change and agreement about which of our activities, particularly those producing greenhouse gases, contribute to climate change. Nevertheless, many climate change models do not offer much of an explanation of the details of the exact mechanism(s) of climate change.

As we discussed in Chapter 1, robustness analysis is a method commonly used in climate science. This sort of analysis involves analyzing a number of incompatible models (i.e., models that make different assumptions about the phenomena being modeled) in order to come up with predictions. In many cases, robustness analysis leads to predictions that are considerably more accurate than those provided by a single model. The reason for this is that in robustness analysis scientists pay particular attention to overlapping predictions of various conflicting models. When models that make use of different assumptions yield the same predictions, those predictions are more likely to be accurate than the predictions offered by a single model or even a set of models that make use of shared assumptions. However, this sort of combined climate model doesn't really explain what is going on because the constituent models disagree with one another. Similarly, we can use statistical data to make predictions without having a corresponding explanation. Take, for instance, the fact that, as reported by the Cancer Treatment Centers of America, it is estimated that about one-third of all women will be diagnosed with some form of cancer in their lifetimes and half of all men will be diagnosed with some form of cancer. Given this fact, we can make fairly accurate predictions about random samplings of men and women. If we look at a group of 100 men and 100 women, we can predict that around 50 of the men and 33 of the women will be diagnosed with cancer in their lifetimes. We can make this prediction without having an explanation for why this is in general, or any explanation in the case of a particular man or woman for why they developed cancer. Being in a position to make a good prediction simply doesn't seem to require having a good explanation.

Hypotheses that offer good explanations tend to allow us to make accurate predictions. But this isn't true in all cases. There are important instances where we have very good explanations but can't use our knowledge of the relevant dependence relations to make comparably good predictions. Let's consider just two familiar examples. First, consider things like earthquakes or the weather more generally. Seismologists can explain the occurrence of an earthquake quite well after one happens. They can describe the mechanism and conditions that resulted in the earthquake. Similarly, after a tornado has occurred meteorologists can explain how the particular atmospheric conditions resulted in the tornado. Unfortunately, in both cases despite having good explanations after the fact, it is extremely difficult to provide accurate predictions before earthquakes or tornadoes occur. At best, seismologists might give a range of dates for when an earthquake is likely to occur and meteorologists might predict, very shortly before the event, that conditions are ripe for tornadoes. Second, think about the development of a particular trait in a given population. Evolution provides an excellent explanation of why a given trait became prevalent in a given population. Take, for instance, the coloring of a particular species of moth. Roughly, the explanation of this is that a random mutation occurred in the species at some time in the past, and individuals with that mutation had a selective advantage (i.e., they survived and produced offspring to a greater extent than those without the mutation). Over time those with the mutation continued to produce more offspring than those without. Knowing this information provides a clear explanation of the predominance of the particular coloring of the moths. But this doesn't seem to put us in a position to make accurate predictions. After all, we know that random mutations will occur, but we don't know whether those mutations will prove advantageous. Furthermore, once a mutation occurs, we cannot know in advance whether the respective trait(s) will be selected in the particular environment in which it arises; or if it were to be selected in that environment, we cannot always know in advance whether or not that particular environment will change in important ways. Hence, before the mutation in question arises in members of the population, we couldn't predict that it would arise or even knowing that it did arise we couldn't predict that it would have the effect that it did. Thus, we might possess a good explanation of why a trait evolved in a particular population without being in a position to accurately predict that it would do so.

Why Does Explanation and Prediction Come Apart?

Sometimes the hypotheses and theories that produce the best explanations don't provide the most accurate predictions, and sometimes the hypotheses and theories that produce the most accurate predictions don't provide the best explanations. Why is it that explanation and prediction come apart at times? This happens because when we are developing theories and hypotheses, or just trying to understand the world around us in general, we run up against two significant challenges. The first is simply that our world is complex. The second is that we are limited in our ability to understand the world for a variety of reasons. Let's take a look at each of these challenges.

To begin, the world that we seek to explain and make predictions about is complex. Philosopher Angela Potochnik helpfully identified four specific dimensions along which the world is complex. First, there is simply a tremendous number of different phenomena in our world. Second, when we consider any phenomenon that we might want to explain, or make predictions about, we find that there are many factors that influence the occurrence or features of the phenomenon. Third, factors that influence the phenomenon that we are examining are often importantly different from the things that influence similar phenomena. Fourth, the relationship that a particular influence bears to a phenomenon of interest can itself be quite complex. In order to get a sense of just how much complexity we are dealing with in explaining the world, let's take a look at an illustration of some of the causal influences on the COVID-19 pandemic (Figure 4.1).

The exact details of this figure are not important for our purposes. What matters is what it helps reveal about complexity. In this figure each node represents a particular factor that has been identified as having a causal influence on the COVID-19 pandemic. Each of the lines between the various nodes represents interactions between those influences. In fact, the causal story of the COVID-19 pandemic is actually significantly more complex than this figure makes it seem because the figure only represents *some* of the causal influences on the COVID-19 pandemic. And, for each of the nodes in this figure we could illustrate the causal influences on it with a similarly complex figure. Consequently, we can see that the COVID-19 pandemic is a vastly complex phenomenon.

One might argue that we do not always (or even often) face this kind of complexity though. After all, the COVID-19 pandemic is a global phenomenon, but many of the phenomena we seek to explain and make predictions about are much more local – the effect of this treatment on a particular group of people, the weather in this particular city, and so on. Hence, one might think that the kind of complexity exhibited by the COVID-19 pandemic is rare. Unfortunately, this isn't so. We face enormous complexity all of the time. Let's consider a figure depicting some of the causal influences on obesity (Figure 4.2).

Like the previous figure, each node in Figure 4.2 represents a causal influence on obesity, and each of the lines represents interactions between those causal influences. And, just as with Figure 4.1, each of the nodes in this figure

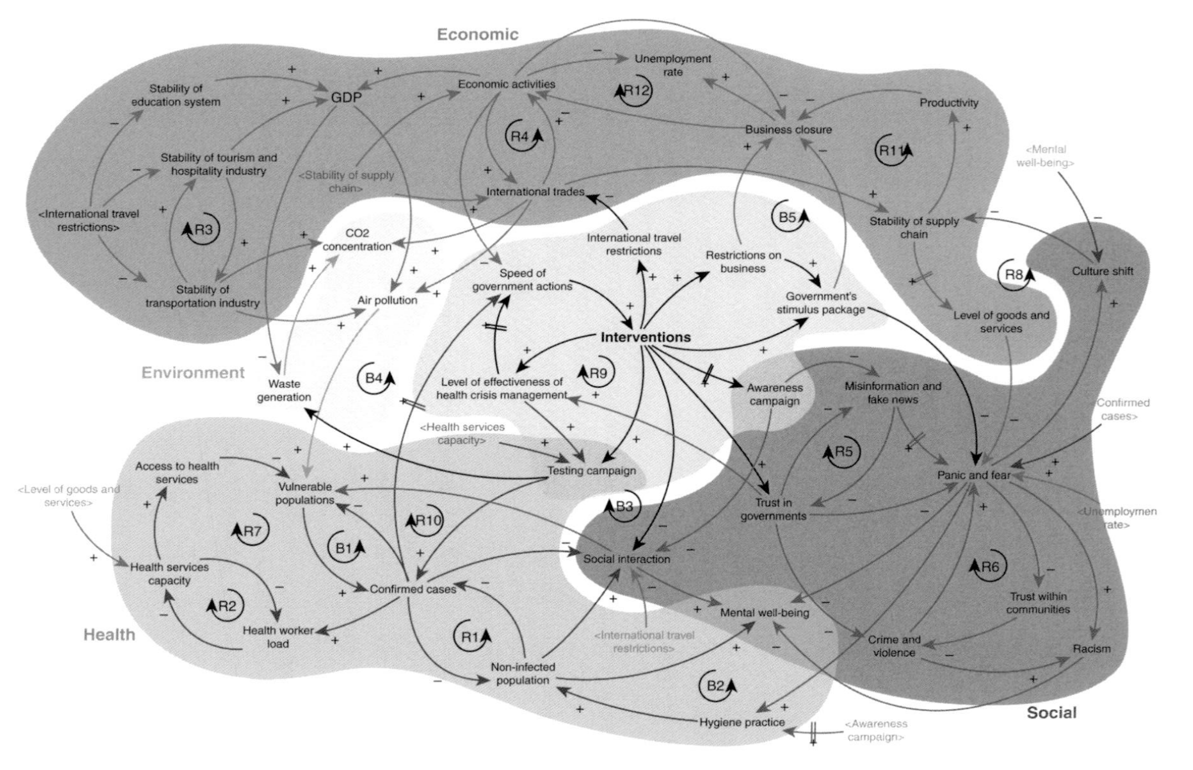

Figure 4.1 Diagram of the causal complexity of the COVID-19 pandemic.

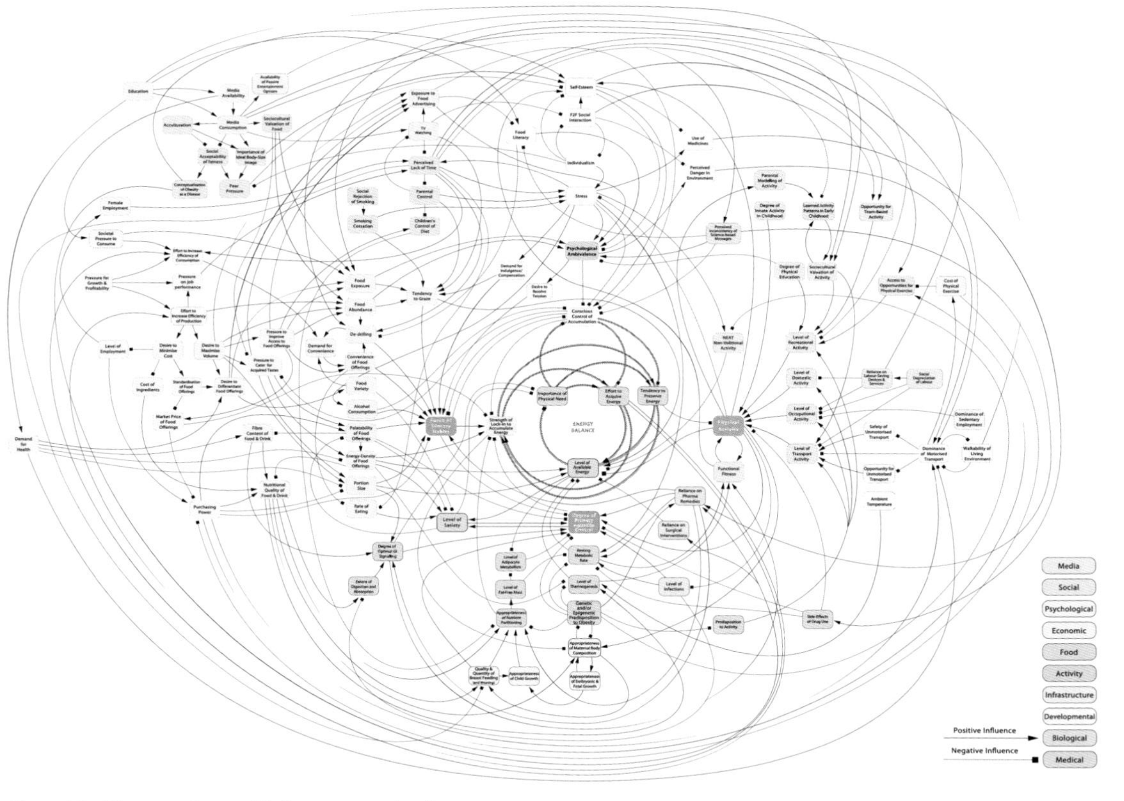

Figure 4.2 Diagram of causal influences on obesity.

represents a phenomenon that is itself subject to a network of causal influences. It is plain to see: even seemingly simple phenomena, such as obesity, are in fact quite complex. Similar figures could be constructed for just about any of the interesting phenomena we wish to explain or make predictions concerning. Complexity surrounds us.

Now, let's think about the second challenge: our limitations. It is undeniable that we are limited in various ways. We have a limited amount of time to devote to studying various aspects of the world. We are limited in terms of our material resources. We are also cognitively limited – we make mistakes, we are subject to biases, and there are simply limits to what we can perceive and comprehend. Admittedly, we have various ways of mitigating our limitations to some extent. We make use of computers, we collaborate with others, we run the same experiment multiple times, and so on. Despite these measures, which certainly do help us overcome a number of our limitations, we are still limited in important ways. (This is something that we will come back to in later chapters.)

In the face of our limitations and the complexity of the world, we construct scientific models and theories. Models in science are like the models we encounter anywhere else; they are representations. Essentially, we use scientific models to represent features of the world in such a way that we can apply scientific theories to those features in order to generate scientific explanations or predictions about the relevant phenomena. To illustrate this, let's consider an everyday example of a model. A map of the New York City subway system represents specific aspects of a target system (the subway system), the locations of specific train stops, the directions various trains travel, and so on. The subway map does not represent every aspect of the subway system, though. For example, it doesn't represent the number of wheels each subway train has, the specific features of the trains' engines, the number of bricks in a given platform, or a host of other things. However, the map does represent the relevant features of the subway system for specific purposes – in particular, the purpose of using the subway trains to get to various locations in the city. Scientific models work in the same way. Such models do not represent every aspect of the systems they represent. Instead, scientific models represent particular aspects in a way that facilitates applying specific scientific theories to generate scientific explanations and predictions about the phenomena that occur in the system represented by that model. In order to do this the model has to leave some things out and sometimes idealize other factors (we'll return to this issue in Chapter 7). As a result, some models and hypotheses are well-suited to providing good explanations; others aren't but are well-suited to generating predictions.

The complexity of the world and our limitations make it so that many times hypotheses and models do not at the same time generate the best scientific explanations and the best scientific predictions. Complexity coupled with our limitations means that we are not typically in a position where we have a complete explanation and so complete understanding of a particular phenomenon. Therefore, we are always working with hypotheses and models that only provide a partial grasp of why and how a particular phenomenon occurs. In such situations, our particular aims will dictate which sort of model or hypothesis is most useful. If we seek to explain, we will pick one kind of model/hypothesis; if we seek to draw predictions, we will pick another kind of model/hypothesis.

How Explanation and Prediction Are Related

Despite the fact that scientific explanations and scientific predictions are distinct, and we often can't have the best of both worlds, the two are very closely related and often work in conjunction with one another. As philosopher Heather Douglas has helpfully explained, "the relation between explanation and prediction is a tight, functional one: explanations provide the cognitive path to predictions, which then serve to test and refine explanations." The best scientific explanations help us understand phenomena in the world because they give us information about dependence relations. Once we know the relevant information about dependence relations, we can make scientific predictions. For instance, once we recognize that certain mutations cause cystic fibrosis, we have an explanation for why people with these mutations have cystic fibrosis whereas those without them do not. With our knowledge of this dependence relation in hand, we might predict that other individuals who have cystic fibrosis might also have these mutations. When we discover that these people do not have the predicted mutations, this will lead us to refine our explanatory hypothesis. In fact, it turns out that more than

2,000 mutations have been identified as being related to cystic fibrosis in humans. But the general point holds – knowledge of dependence relations (i.e., good explanations) often allows us to make useful predictions. Scientific explanation and scientific prediction are not the same, but they are very closely related central aims of science and both are critical in our quest to explain the world.

5 Evaluating Explanations

Better and Worse Explanations

Even though explanation plays a central role in science, it is not enough to simply come up with explanations. Scientists (and everyone else) must also *evaluate* explanations. After all, it's clear that not every explanation is a good one, as well as that some explanations are better than others. For example, evolutionary theory provides a much better explanation of the diversity of life than, say, the hypothesis that all organisms appeared at the same time in their present form. But what makes one explanation better than another? Relatedly, how can we tell which of a set of competing hypotheses provides the best explanation?

Our focus in this chapter is to investigate how we evaluate explanations. In general, we evaluate explanations by looking at their theoretical virtues. So, when we are comparing the explanations offered by two hypotheses, we can determine the better explanation by figuring out which explanation has the best complement of theoretical virtues. To achieve this, we need to know not only what sorts of things count as theoretical virtues, but also whether we can be confident that we're any good at spotting such virtues, and whether theoretical virtues are really something that should guide our theorizing at all.

When it comes to theoretical virtues, there isn't a consensus as to exactly how many there are or what exactly we should call each of them. Nevertheless, we can approach the task of exploring theoretical virtues by limiting our discussion to only the most commonly accepted ones, and we will refer to them by commonly accepted names. Let's start by taking a close look at several theoretical virtues.

Explanatory and Predictive Power

Let's begin with perhaps one of the most widely accepted (and most straightforward) theoretical virtues: explanatory power. The more data – particularly of different types and from independent sources – that a hypothesis explains, the more explanatory power it has. Its vast explanatory power is one of the key theoretical virtues touted in favor of evolutionary theory. For instance, natural selection explains the vast array of different adaptations in organisms, why some traits come to be found in a population and others die out, how bacteria become resistant to antibiotics, and much more. As we saw in Chapter 1, Darwin himself, in *The Origin of the Species*, appealed to its explanatory power as significant evidence in support of natural selection.

Predictive power is a theoretical virtue that is as widely, if not even more so, accepted as a theoretical virtue as explanatory power. In fact, some think that if a hypothesis doesn't generate testable predictions, it is not really scientific at all – regardless of how explanatory the hypothesis is. This sort of concern lies at the heart of the debate concerning whether string theory is genuinely scientific or not, for example. That said, it would be a mistake to think that providing accurate predictions is all that matters when it comes to the quality of an explanation. After all, in the previous chapter we saw that explanation and prediction are not merely two sides of the same coin; they differ in important ways. Sometimes the most explanatory hypothesis will not be the one that yields the most accurate predictions, and vice versa. In general, if we are comparing two hypotheses that explain various data, the fact that one of them has more predictive power is certainly a point in its favor. Importantly, making predictions isn't in and of itself all that important for the quality of an explanation. For instance, a hypothesis that makes only inaccurate predictions is no better than an explanation that makes no predictions at all, and in some ways it is worse. Consider two hypotheses, H_1 and H_2. H_1 makes no predictions, while H_2 makes many predictions but they've all been shown to be mistaken. In such a case, we may have reason to question whether H_1 is genuinely scientific, but we don't have evidence for thinking that it is false. However, while H_2's credentials as scientific don't necessarily seem to be in question, we have good reason to think that it is false. The lesson here is that genuine predictive power involves not just making testable predictions, but also accurate ones. And, obviously, a theory or hypothesis that has made accurate predictions is clearly superior to one that hasn't (all else being equal).

When it comes to actual scientific practices, it is plausible to think, as philosopher Michael Strevens suggests, that the rules of scientific debate allow "nothing but matters of explanatory power, nothing but a theory's ability to account for the observable, to determine the course of scientific argument." For these reasons, explanatory power and predictive power are sometimes referred to as "empirical virtues" rather than theoretical virtues. The thought is that explanatory power and predictive power each concern not only a hypothesis's own structure or intrinsic features, but also its relationship to the data. Hence, even those who are skeptical of other theoretical virtues are apt to think that differences in explanatory power and predictive power make one explanation better than another. As we will see, however, it's not clear that there is a sharp distinction to be made between so-called empirical virtues and the other theoretical virtues. For now, we can simply note that explanatory power and predictive power are key components of evaluating explanations.

Conservatism

Another theoretical virtue that is commonly used to evaluate explanations is conservatism. As the name suggests, the idea behind this virtue is that all things being equal it is best to *conserve* what we already accept. In particular, the hypotheses that require the fewest changes to our already accepted theories are preferable to those that require more changes, at least when these hypotheses are otherwise on a par. Conservatism is sometimes dismissed as merely something that makes things easier for us without giving us a reason to think that conservative hypotheses are more likely to be true or even better explanations than less conservative rivals.

First, we can, and should, readily acknowledge that given limited resources and our own cognitive limitations, as we discussed in the previous chapter, we have practical reasons to prefer hypotheses that require minimal changes to what we already accept. But, of course, the primary concern still remains: why think that the conservativeness of a hypothesis gives us reason to think that the hypothesis is true? The reason is that we typically have good evidence in support of the theories and hypotheses that we already accept. This is especially so when we consider well-established scientific theories such as evolution or general relativity. Since we have strong evidence in support of these theories, hypotheses that are less conservative – that is, hypotheses whose acceptance would require us to revise or give up such well-established theories – go against this evidence. Thus, less conservative hypotheses face strong opposing evidence that more conservative hypotheses do not. Of course, a new hypothesis or new data could come along that leads us to revise or give up on well-established theories. All of science is revisable – no scientific theory is sacrosanct! Nevertheless, if we have a choice between two otherwise equal hypotheses, one of which requires us to abandon a well-established theory whereas the other doesn't, we have strong reason to favor the more conservative hypothesis because it doesn't have to overcome the evidence supporting the established theory like the more revisionary hypothesis does.

Second, while it is true that the two hypotheses (a conservative one and a revisionary one) might explain the same data, the fact that one is more conservative means that it is more explanatory. The reason is that the more conservative hypothesis fits better into the explanatory picture provided by the well-established theories that it is compatible with – the very theories with which the revisionary theory is in conflict. As a result, the overall explanatory system that results from accepting a conservative hypothesis is richer and more coherent than the competing system that results from accepting the revisionary hypothesis (assuming that all other things are equal). Now, it could be that a revisionary hypothesis suggests a new theory that could replace a well-established theory and lead to an even richer explanatory framework. In such a case, the conservativeness of its rivals may be outweighed by the quality of the new explanatory framework of which the less conservative hypothesis is a part. Nonetheless, in general more conservative hypotheses are explanatorily superior to less conservative hypotheses (all other things being equal).

Simplicity

We now turn our attention to what is, perhaps ironically, one of the more complex of the theoretical virtues: simplicity. What makes simplicity so complex is that there are a number of different things we might look at. For instance, sometimes simplicity is understood in terms of the number of individual things a hypothesis posits, but other times simplicity is a matter of the number of *kinds* of things rather than things themselves. Imagine that we are considering two competing hypotheses, each of which explains the occurrence of a particular disease, such as bone cancer. One hypothesis says that bone cancer is caused by a particular mutation of a single gene. The other says that bone cancer is caused by someone having particular mutations of three separate genes. In this case, the first hypothesis is simpler than the second. Although they both posit the same *kind* of cause, the first posits fewer individuals of that kind. Now compare the second hypothesis to a third, which, like the first, says that bone cancer is caused by a particular mutation of a single gene but adds to this that the mutation only leads to cancer in the presence of significant amounts of radiation. It is not clear whether the second hypothesis or the third is simpler. On the one hand, the second hypothesis posits more individual causes of a particular type (it says that bone cancer arises when one has particular mutations of three different genes, whereas the third hypothesis only appeals to one mutation and the presence of radiation). On the other hand, the third hypothesis posits more kinds of causes (it posits both genetic mutation and radiation, whereas the second only posits mutations). Each hypothesis is simpler than the other in one sense but more complex in another. This can make it difficult to determine whether a given hypothesis is the simplest all things considered.

Despite the complexities of simplicity, many of the most famous scientists in history have appealed to simplicity as a key theoretical virtue. To take two examples: Isaac Newton defended his "Rules for the Study of Natural Philosophy" in his magnum opus, *The Principia*, by appealing to simplicity. As he said, "nature is simple and does not indulge in the luxury of superfluous causes" and "nature is always simple and ever consonant with itself." Similarly, Albert Einstein claimed that "our experience hitherto justifies us in trusting that nature is the realization of the simplest [theory] that is mathematically conceivable." Later, writing with Leopold Infeld, Einstein said, "without belief in the inner harmony [and simplicity] of our world there could be no science." Unfortunately, neither Newton nor Einstein actually provided us with a precise definition of simplicity. It was, apparently, something that both of them took to be fairly obvious when encountered. Importantly, they were both committed to simplicity as a theoretical virtue that was of central importance when evaluating theories. This is something that many scientists continue to accept today. For instance, when biologist George C. Williams argued in favor of genes as the unit of selection rather than groups or organisms, he did so at least in part on the basis of the simplicity of his hypothesis compared to its rivals. Williams contended that the group selection hypothesis has to appeal to two distinct processes, whereas the gene-selection hypothesis only has to appeal to one to explain the same data. When it comes to group selection, explaining a trait's being selected for involves appealing to both within-group selection and between-group selection. The gene-selection hypothesis can account for a trait's being selected by appealing to just the process of individual selection.

But, if it is not clear which sort of simplicity matters and hypotheses can be simpler in some respects and more complex in others, how can we appeal to simplicity when evaluating theories? First, as we noted above, many scientists (such as Newton and Einstein) seem perfectly able to tell which hypothesis is simpler without being able to give a general account of *simplicity*. This shouldn't surprise us. We are often able to determine something when we encounter it without giving, or perhaps even being able to give, a precise definition or description of it. Second, there are different kinds of simplicity because there are different aims, methods, standards, and so on in different scientific contexts. Plausibly, which form of simplicity is apt for comparing hypotheses will depend upon the particular context and scientific community in which the comparison is being made. However, this doesn't make simplicity merely a matter of taste. It will be an objective fact whether or not a particular hypothesis is simpler than another when it comes to specific kinds of simplicity. What will vary from context to context is which form of simplicity is most relevant to evaluating the theory. Thus, the variability of kinds of simplicity doesn't seem to pose a problem for evaluating hypotheses according to their possession of this virtue.

At this point, one might be tempted to grant that perhaps we can use simplicity when evaluating theories, but wonder if we *should*. Importantly, one might worry that simplicity is merely an aesthetic property of hypotheses rather than a property that gives us reason to think that they are true. In the final section of this chapter we'll meet this challenge head-on. But for now, we can simply note that in many cases the simplicity of a particular hypothesis has been argued to allow it to better account for the data. Copernicus argued in this way to support the heliocentric model of the solar system. And, as we've mentioned, both Newton and Einstein held simplicity in high regard. Of course, we don't simply want to appeal to authority on this issue, but then again, it would be a mistake to dismiss out of hand the opinions of scientists who are credited with developing some of the most successful scientific theories ever produced.

Before setting this concern aside until the end of the chapter, it's worth thinking of one commonsensical reason for preferring simpler theories to more complex ones – aside from the practical reason of simpler theories being easier to work with: the more complex a theory is, the more ways it can be wrong. For example, if a hypothesis only appeals to the existence of one type of causal mechanism and another posits that type plus another type, there are more ways for the second hypothesis to be wrong compared to the first one. Since the first is only committed to the existence of one mechanism, roughly, it is mistaken only if that mechanism doesn't exist or doesn't function in the way the hypothesis says. The second hypothesis, though, is wrong if either the mechanism mentioned by both hypotheses fails to exist or doesn't work as they say, *or* if the other mechanism that it alone posits fails to exist or doesn't work as it says. Put simply, more complex hypotheses are committed to more claims than simpler hypotheses, and so there are more ways that they can go wrong.

Beauty

Sometimes simplicity is touted as a feature that makes hypotheses beautiful. While this is so, we should be careful not to conflate the two because simplicity and beauty are distinct theoretical virtues, despite their close relationship. There's a reason why aesthetics is a discipline in its own right and why artists as well as critics have a hard time saying exactly what makes something beautiful. Hence, we will have to content ourselves with saying a bit about some of the common features that are frequently said to make a hypothesis beautiful, discussing examples in science where beauty is appealed to in support of hypotheses, and offering a brief argument for why we should think that beauty, although clearly an aesthetic virtue, should also be considered a theoretical virtue.

We've already said that simplicity is often characterized as a feature that makes (or enhances) a hypothesis's beauty. Since it's difficult to say exactly what simplicity is, it's hard to say with any precision what exactly about simplicity can make a hypothesis beautiful. But, in addition to simplicity, another feature that is often said to make a hypothesis beautiful is balance or symmetry. Roughly, symmetry refers to structures that remain the same despite transformations. For example, a physical object has symmetry when it has features that are roughly the same on both sides. This allows the object to look basically the same in spite of being "transformed" by being rotated or moved in various ways. Such symmetry can occur in hypotheses and formulas when replacing one input with another leaves things virtually unchanged. Perhaps an easy way to see this is to consider the double-helix structure of DNA. This structure exhibits symmetry in that the two strands match up with one another. In fact, biologist James D. Watson (of the Watson and Crick model of the structure of DNA) said of Rosalind Franklin (an X-ray crystallographer who made significant contributions to the discovery of the structure of DNA) that she "accepted the fact that the structure was too pretty not to be true."

In addition to simplicity and symmetry, another feature that is often said to make a hypothesis beautiful is inevitability or completeness. Hypotheses have this feature when their parts fit together so well that it seems that every component must be there – none could be removed without spoiling the whole. As Albert Einstein put the point when speaking of his theory of general relativity, "The chief attraction of the theory lies in its logical completeness. If a single one of the conclusions drawn from it proves wrong, it must be given up; to modify it without destroying the whole structure seems to be impossible." Although this is far from exhaustive, recognizing the role of simplicity, symmetry, and inevitability/completeness gives us a fairly good grasp of what it takes for a hypothesis to be beautiful.

There are many more examples of scientists appealing to beauty when theorizing. Let's consider at least a few more examples to highlight the role that the search for beauty plays in science. James Clerk Maxwell seems to have been guided by considerations of beauty when formulating his famous equations of classical electrodynamics. As physicist Roger Penrose has noted, the "symmetry generated [by these equations] must have played an important role for Maxwell in his completion of these equations." The physicist Paul Dirac claimed of the general theory of relativity that anyone with understanding of how nature works "must feel that a theory with the beauty and elegance of Einstein's theory *has* to be substantially correct." Physicist Murray Gell-Mann's eightfold way and his discovery of quarks was led by a search for beauty in theories. We could go on, but this is enough for the point to be sufficiently clear: scientists take the beauty of a hypothesis seriously as a sign that the hypothesis is correct. It is worth noting that in general features that contribute to the beauty of a theory such as its offering a visualization of phenomena are prized in the biological and social sciences just as they are in physics. Beauty plays an important role in scientific theorizing in general. Physicist Steven Weinberg goes so far as to say that "we would not accept any theory as final unless it were beautiful."

But why are scientists correct in allowing considerations of beauty to guide their theorizing? Simply put, the reason is that *it works*. As Michael Strevens has noted, when we look at the history of science, "the successes [of being guided by beauty in theorizing] are too important, too frequent, and too dazzling to ignore." We have already mentioned a number of such successes. However, we can take note of a more general phenomenon that further strengthens the case for treating beauty as a guide to truth. Physicists have noted that mathematicians tend to develop theories with the aim of discovering those with beautiful structures. Despite mathematicians being solely focused on beauty, physicists quite often later discover that the formal structures mathematicians developed because of their beauty are immensely helpful to physicists' attempts to understand natural phenomena. As Steven Weinberg has explained the phenomenon, "physicists generally find the ability of mathematicians to anticipate the mathematics needed in the theories of physicists quite uncanny. It is as if Neil Armstrong in 1969 when he first set foot on the surface of the moon had found in the lunar dust the footsteps of Jules Verne." It seems that we have two explanations for this uncanny phenomenon: it is either random chance, or beauty is indeed a fairly reliable guide to truth. Just as it would be unreasonable for Neil Armstrong in the imagined scenario to conclude that Verne's footsteps are there by chance, it would be unreasonable for us to conclude that the link between beautiful hypotheses and these successes is merely a matter of chance.

Which Explanation Is Best?

We've now looked at some of the primary theoretical virtues that contribute to the quality of a given explanation. It's natural now to ask: How can we tell which explanation is the most virtuous? Or, put another way, what makes one explanation better than another? A good answer to this question is that the best explanation is the one that would, if true, provide the most understanding of the phenomena that it seeks to explain. Essentially, the explanation that has the best combination of theoretical virtues is the one that is best in this sense. Obviously, this leads to yet another question: How can we tell which explanation has the optimal mixture of the various virtues? Unfortunately, this is a very difficult question to answer with any precision. This is especially so because some virtues seem to pull us in opposite directions. For instance, we can increase explanatory power by adding more features and auxiliary hypotheses to a given hypothesis, but, of course, doing so makes the resulting hypothesis more complex than it was. Such added complexity might also detract from the beauty of the hypothesis as well. When we have to choose between improving with respect to one theoretical virtue at the expense of another theoretical virtue, how do we choose? Much will depend upon the particular context in which we find ourselves.

Even without clear answers as to how we best balance theoretical virtues, we can readily tell (in at least some cases) when a hypothesis/theory is a very good explanation and when it is better than its rivals. A prime example is evolution. Evolutionary theory has a tremendous amount of explanatory power and predictive power. For example, evolutionary theory explains why we see so many fundamental similarities between the DNA and cellular structures of different organisms. Evolutionary theory also makes numerous successful predictions. One of the first came from Darwin himself. After examining a particular species of orchid found in Madagascar, Darwin predicted that a species of moth would be found on Madagascar that had a very long tongue so that it could reach to the bottom of the orchid's foot-long nectary. His basis? Given evolutionary theory, it is very probable that such a moth existed because the orchids and moths in Madagascar evolved together. Although his prediction was at first met with derision by some entomologists, it was later empirically confirmed. Despite being quite radical when it was initially proposed, evolution is now quite conservative. It doesn't conflict with any of our best confirmed scientific theories. In spite of its vast explanatory power and its numerous predictions, evolutionary theory is surprisingly simple and, on the whole, it is arguably a beautiful theory. On top of all of this, there are no serious contenders for a genuine explanatory rival of evolutionary theory. As a result, in this case it is very easy to see that evolution is the best explanation when it comes to the existence of the various species that we find in the world.

One might worry that while it is easy to see that a theory that has already been widely accepted is the best explanation, it is not so clear beforehand. Obviously, when there are multiple rival hypotheses that explain the same data, that makes it more difficult to determine which hypothesis provides the best explanation. Nonetheless, this doesn't mean that we can't make such determinations. Nor does it mean that we have reasons for thinking that scientists aren't good at spotting the best explanation in general. Now, we have already recognized above that sometimes which theoretical virtues are considered most important is a contextual matter. So, in some instances, simplicity considerations might give way to considerations of explanatory power, say. In practice, scientists can learn how to make judgments concerning which explanation is best in their particular context by way of studying exemplars. Learning of examples of excellent explanations in a given field helps equip scientists to identify the best explanations in the context of their own research in that field. Furthermore, scientists (sometimes with the help of philosophers of science and others) can also work to get clearer on the nature of individual theoretical virtues to make the process even more precise.

In order to fully put the concern about identifying the best explanation to rest, let's move from discussing how scientists can spot the best explanation to discussing children. There are numerous studies showing that not only do children make judgments of explanatory quality, but also that they are quite good at it. Unsurprisingly, psychologists have found that early in development children are curious and tend to seek explanations for the things they observe.

What is perhaps surprising is that even very young children appear to have fairly strong intuitions about what makes for good explanations. For example, children as young as six years old show a preference for simpler explanations. In fact, there is even some evidence that *infants* make use of explanatory evaluations in some contexts. Importantly, the intuitions that children (and adults) have with respect to explanatory quality are both strong and systematic. When we think about this it shouldn't really strike us as all that surprising. After all, as psychologist Tania Lombrozo has aptly noted, "Children, adults and scientists alike confront the world with a common question: why?" It is for this reason that she has pointed out that developing and judging explanations is something that we do all the time in our attempts to understand the world. Plausibly, evaluating explanations is a key aspect of how we come to understand the world around us. Since we all do this all the time and we start doing it as young children, we have a lot of practice with it! In light of the facts that evaluating explanations is ubiquitous and that even young children are fairly proficient at it, our being unable to spell out precisely how to adjudicate the relative importance of conflicting theoretical virtues seems to be of little concern.

Is the World Simple/Beautiful?

Why should we think that simplicity or beauty or any other theoretical virtue is a guide to truth rather than something that we simply prefer for aesthetic or pragmatic reasons? Why should we think that our world is simple? The challenge we face here is that of showing that theoretical virtues can really make it so that we have good *epistemic* reasons (that is, reasons related to truth or knowledge) for preferring more virtuous hypotheses to less virtuous ones. For instance, let's think about the Alvarez hypothesis, which holds that the Cretaceous–Paleogene (K–Pg) extinction was caused by an asteroid striking the Earth. The Alvarez hypothesis possesses a number of theoretical virtues, but rather than enumerate them, let's simply assume that it does very well with respect to each of the theoretical virtues we have discussed. Let's also assume, at least for the sake of simplicity, that there are no plausible rivals to the Alvarez hypothesis. Does the fact that the Alvarez hypothesis possesses these theoretical virtues give us good reason to believe that it is true or even likely to be true?

One route to responding to this challenge is to attempt to argue on theoretical grounds that individual theoretical virtues, such as beauty, are reliable guides to the truth, or similarly to argue that theoretical virtues in general are truth-conducive. So, without worrying about the particular details of the Alvarez hypothesis, we might try to show that there are grounds for thinking that each particular theoretical virtue is a guide to truth or that taken as a whole such virtues are apt to belong to true hypotheses more often than they do to false hypotheses. There have been a number of attempts of this sort, and we actually illustrated some of the many successes of various virtues above. Rather than enter into this primarily philosophical debate, however, we will take another route here. Our approach doesn't require in-depth philosophical analysis or a deep dive into the theoretical literature surrounding this particular challenge. Instead, we will simply take note of some of the successes that have resulted from preferring hypotheses because of their theoretical virtues.

Recall from above the many instances where appeals to simplicity or beauty played an integral role in enormously important scientific successes. Also recall that theories are accepted because of the vast superiority of their explanations when compared to rivals, or because they are excellent explanations with no serious rivals. Nevertheless, there may be instances where accepting a hypothesis because it provides the best explanation leads us astray – such as when the mathematician Urbain Le Verrier suggested that there is a small planet between the sun and Mercury because the presence of this planet, which he called "Vulcan," best explained peculiarities in Mercury's observed orbit. There is no such planet, but at the time the existence of Vulcan best explained the observations of Mercury. So, accepting hypotheses because they offer the best explanations doesn't always work out. But no scientific reasoning is infallible. Our claims to scientific knowledge are always held tentatively, and we should be ready to revise in the light of new evidence.

Fortunately, failures like Le Verrier's postulation of Vulcan tend to be corrected later because a more virtuous theory comes along. Although Le Verrier's postulation of Vulcan best explained Mercury's orbit *assuming* Newton's theories, a better explanation came along. General relativity replaced Newton's theories and provided a better explanation of Mercury's orbit without positing a nonexistent planet. It is theoretical virtues that have led us to the more accurate picture of the world provided by general relativity.

To paraphrase philosopher J. D. Trout, we know that following theoretical virtues works, so the ultimate purpose of raising largely philosophical doubts about the link between such virtues and the truth is unclear. At the end of the day, whether or not we have reason to believe that the world is simple, or beautiful, or whatever, we have good reason to think that scientists have made, and are continuing to make, tremendous progress in understanding the world by adopting hypotheses that best exemplify theoretical virtues. That is enough to warrant trusting that such virtues are good indicators of the truth.

6 Explanatory Quality and Felt Understanding

Explanation Aims at Understanding

A general way of appreciating some of the main ideas of the previous chapter is to recognize that explanations aim at providing understanding. Scientists and philosophers agree that understanding is a (if not *the*) primary epistemic goal of scientific inquiry. Both explanation and prediction tend to be closely related to understanding. We want explanations in science because we want to understand why the world is as it is and how things happen. And, once we understand various phenomena, we can make accurate predictions about them. One simple, and widely accepted, way of assessing the quality of a given explanation is to look at the understanding it provides. As philosopher Peter Lipton explained, the explanation that is the best is simply the explanation that, if true, would provide the deepest understanding of the phenomena being explained. That being said, some worry that this way of assessing explanations is problematic because it seems that we might misjudge how well we understand something. This is a genuine concern, and we will address it in this chapter, but before doing so let's first further clarify (recall our brief discussion in Chapter 1) how exactly we should think about understanding.

Kinds of Understanding

What exactly is understanding? We have to exercise some caution here for at least two reasons. The first is simply that like any other important concept there is some controversy concerning exactly how best to construe it. The second reason, as we saw in Chapter 1, is just that there are a variety of ways that the term "understand" is used. For example, sometimes we use "understand" as a means of not fully committing to what we say. If you don't want to flat-out accuse someone of being dishonest, you might say something like, "I understand that you didn't really mean what you said." At other times we treat "understand" as synonymous with "knowledge." Saying something like "I understand that there is a meeting today" is just another way of saying that you know that there is a meeting today. Neither of these are really the sense of "understanding" that is relevant when it comes to science. Instead, we want to understand why and how things are as they are. For instance, why is it that fish need oxygen to live? How do fish extract oxygen from water? And so on. The sort of understanding that is pertinent to science is understanding of natural phenomena. Consequently, when it is said of you that you understand why fish need oxygen and how they extract it from water, a cognitive achievement has been attributed to you: grasping why fish need oxygen and how they get it from water. Simply put, in this case we are saying that you possess correct explanations of these phenomena.

Let's dig a little deeper. Understanding involves grasping a correct explanation, but how should we understand this grasping? A simple answer is that the relevant grasping of an explanation is just knowledge. Hence, we might be tempted to say that you understand why fish need oxygen and how they get it when you know accurate explanations of these phenomena. This simple answer also seems to provide a fairly straightforward account of how one might have deeper understanding of a topic than someone else. One person has a deeper understanding of, say, biology when she knows more scientific explanations of biological phenomena than someone else.

Although thinking of understanding as just knowledge makes things concise and has the theoretical virtue of simplicity, it doesn't seem to be the full story. To see this, think about someone who knows a subject without really understanding it. Philosopher Catherine Elgin describes such a case with respect to geometry. Elgin asks us to consider the possibility of someone who "knows all the axioms, all the major theorems, and their derivations," but who came to possess this knowledge through rote memorization. In other words, the person accepted these facts on the basis of someone else's testimony and has simply memorized them. Elgin rightly insists that in this situation the person lacks understanding of geometry, even though they know a tremendous number of geometrical facts. What is missing in this case? To make some progress on this question let's first think about what sorts of abilities someone with genuine understanding of geometry has that the memorizer lacks. According to Elgin, the person with genuine understanding "can reason geometrically about new problems, apply geometrical insights in different areas, assess the limits of geometrical reasoning for the task at hand, and so forth." By contrast, the memorizer seems to lack these abilities. As a result, there appears to be significant differences between the person who merely knows geometrical facts and the person who actually understands geometry.

A natural question at this point is what is required for understanding beyond knowledge? The widely accepted answer to this question is that understanding requires "grasping" (understood as not simply knowing accurate explanations), but knowledge does not. What exactly does this grasping amount to, though? Here we run into a difficulty because there have been woefully few well-developed accounts of grasping. What has been said though, and what seems most plausible, is that understanding requires certain abilities that knowledge doesn't. As we saw in the previous paragraph, the reason that we tend to think that the person who simply memorized geometrical facts lacks understanding is precisely because they cannot really put their knowledge to use. By contrast, the person with genuine understanding of geometry can. The person with understanding can see how the facts that they know fit together in various ways and can put this knowledge to use in thinking about a variety of geometrical problems.

Now that we have a firmer grasp on what is required for understanding in general, let's clarify scientific understanding in particular a bit more. Following philosopher Henk de Regt, we can distinguish three senses of understanding relevant to science: the phenomenology of understanding, understanding a theory, and understanding a phenomenon. The phenomenology of understanding is the felt experience of understanding, which we discuss more fully in the sections that follow. Roughly, this is the sort of sense that we have when we take ourselves to understand something. Understanding a theory requires both knowing the details of the theory and being able to put that theory to work in one's reasoning. Returning to our geometry example: the person who has only memorized geometrical facts lacks this sort of understanding of geometry. Why? Admittedly, the person knows the details of geometry, but they lack the ability to put geometrical theory to work in their reasoning. They can't solve novel geometry problems or even appreciate the solutions to such problems. The final sort of understanding is the sort that is most clearly connected to scientific explanation: understanding phenomena. This is the kind of understanding that we typically think of when we talk about scientific understanding. Someone has this kind of understanding when they have an accurate scientific explanation of why or how a particular phenomenon occurs as it does. Depending on the quality of the explanation, the person might have a deeper or shallower understanding of the phenomenon. It is this last sort of understanding that is the primary aim of science.

Putting this together, we can see that scientific understanding is generated when someone understands a scientific theory and is able to use that theory to generate (or at least appreciate) scientific explanations (and predictions) of various phenomena. Such understanding comes in degrees. The more aspects of a particular phenomenon one can correctly explain or predict, the better one understands that phenomenon. This point is something that we naturally latch onto when we talk about people having deeper or better understanding of something. For instance, we might correctly ascribe understanding of genetic drift to both an introductory biology student and a professional biologist. Nevertheless, it would also be correct to claim that the biologist has a deeper understanding of genetic drift than the introductory student. After all, the biologist grasps more dependence relations between genetic drift and other phenomena than the student does. As a result, she can appreciate a wider range of biological explanations related to genetic drift and generate more accurate predictions of the results of genetic drift in various populations than the student can.

At this point, one might wonder where the other concept that de Regt mentions, the phenomenology of understanding, comes into the picture. In many cases when one genuinely has scientific understanding, one will have a phenomenologically recognizable sensation of having understood. This is the sort of "aha" experience that one has when a solution becomes clear or it strikes one that a problem is solved. Although this experience often accompanies genuine understanding, it doesn't always. There may be instances where we *feel* like we understand when we really don't.

Nothing More than Feelings?

There are various reasons why we might think that we understand something when we don't. As J. D. Trout argues, we are prone to numerous cognitive biases that may lead us to feel that we understand – that is, give us an "aha" experience or impression that we have hit upon the right answer – when in fact we are mistaken. Two such biases are hindsight bias and overconfidence bias. Hindsight bias is the bias where one tends to think after something has occurred that one could have predicted it beforehand. For instance, after two evenly matched teams compete, it is common for people to claim that they knew which team was going to win beforehand. Similarly, many people in the financial sector claimed (after the fact, of course) that they knew the 2008 financial crisis was bound to occur. Presumably, in both cases people didn't know before the event that it was going to happen. Furthermore, they weren't in a position to accurately predict what was going to happen beforehand. Quite often our take on things with the benefit of hindsight is clearly a mistake – we had no chance of reliably predicting what was going to happen beforehand, despite thinking that we could have done so after the fact. As noted above, *if* we really do understand a particular phenomenon, then that understanding will often put us in a position to make accurate predictions. We recognize predictive abilities as grounded in understanding, and at times we erroneously think we could have predicted some phenomenon. As a result, we mistakenly think that our merely perceived ability to make predictions indicates understanding.

Overconfidence bias is the bias where we are overconfident of ourselves and our abilities. It seems that this bias is very pervasive. In fact, psychologist Thomas Gilovich has noted that a "large majority of the general public thinks that they are more intelligent, more fair-minded, less prejudiced, and more skilled behind the wheel of an automobile than the average person." An example of the sort of evidence garnered to support the prevalence of overconfidence bias comes from a study of one million US high school students. This study found that "70% [of students] thought they were above average in leadership ability, and only 2% thought they were below average. In terms of ability to get along with others, *all* students thought they were above average, 60% thought they were in the top 10%, and 25% thought they were in the top

1%." Clearly, a number of these students were mistaken – while efforts to encourage students' self-esteem are well-intentioned, it simply cannot be that all students are above average in their ability to get along with others! And students aren't the only ones who suffer from this bias. A survey of college professors found that "94% thought they were better at their jobs than their average colleague." It is this widespread overconfidence that leads us to think that we are better at predicting the future than we are. As a result, psychologist Daniel Kahneman has suggested that overconfidence may well be the most significant cognitive bias that we confront. One of the consequences of overconfidence bias is that we often tend to overestimate our understanding of things. As a result, we may feel that we understand something when we have little to no understanding of what's really going on.

Given the presence of hindsight, overconfidence, and other biases, as well as the fact that such biases can lead us to erroneously feel that we understand when we don't, some might be tempted to follow Trout in concluding that understanding simply isn't important for science. After all, if the feeling of understanding can lead us astray, maybe understanding doesn't really matter. Dismissing the importance of understanding for science on this basis is a mistake for at least three reasons, however.

First, although there are numerous studies showing that hindsight bias and overconfidence bias (among others) are widespread, we must be careful in how we think about these studies. And, we have to avoid the error of reading more into the studies than we should. For example, in the above study it was noted that 70 percent of US high school students think they are above-average leaders. This doesn't mean that the majority of US high school students have inflated views of their leadership ability, though. Even assuming that the results of the study are representative of the entire population of US high school students, it doesn't follow that the *majority* of students overestimate their leadership ability. The reason is simply that a number of students *are* above average. Putting things a bit crudely, around 50 percent of US high school students are above the average. Hence, it is possible that only 20 percent of the students in this study actually overestimate their abilities. It's consistent with the findings that the other 50 percent who think that they are above-average leaders really are above average. Even in the worst case, where all the students rated themselves above average when it comes to getting along with others, at most this shows that 50 percent of the students are wrong about this. It may be, as psychologist Gerd Gigerenzer has argued, that studies of this sort, those that seek to show that we are prone to systematic cognitive errors, may not actually show what they are commonly taken to establish. In this case it may be that the study fails to show that students are making significant errors about their abilities.

More generally, Gigerenzer has insisted that many of these studies ignore important pragmatic and context-dependent features that play a key role in human reasoning. Of course, it's also possible in a case like the study at hand that vagueness in the question (what does "ability to get along well with others" really mean and how do we measure it?) could account for why so many students rate themselves above average. It may be that students interpret this question in different ways, so that many of them are not really overestimating their abilities, given their understanding of what the question is asking them. The lesson here is that we must be careful not to mistake judgments under uncertainty for cognitive errors. Obviously, this is not to say that hindsight bias, overconfidence bias, and other biases are not serious or that we shouldn't work to avoid falling prey to them. But it is to say that we need to be careful to not overestimate the prevalence of such biases or what they show us about the reliability of the feeling of understanding. As Kahneman has said, discovering that we occasionally run afoul of biases or other cognitive errors does not "denigrate human intelligence, any more than the attention to diseases in medical texts denies good health. Most of us are healthy most of the time, and most of our judgments and actions are appropriate most of the time."

The second reason that we shouldn't take our occasional unreliability when it comes to the feeling of understanding to undermine the importance of understanding in science is the fact that it is *occasional*. The most that the empirical studies of hindsight bias, overconfidence bias, and other cognitive errors show is that we *sometimes* feel we understand when we don't. They don't show that our feeling of understanding is mostly misleading when it comes to having genuine understanding. Furthermore, our awareness of the prevalence of biases can help us to be more circumscribed in our trust of the feeling of understanding. Hence, such a feeling is not a perfectly reliable guide as to when we understand, but it may, nevertheless, often be an important indicator (perhaps one among many) of understanding.

The third, and most important, reason to not follow Trout in thinking that understanding is unimportant for science is simply that, as we have seen above, the phenomenal feeling of understanding isn't what we are after in science anyway. In other words, those who claim (correctly) that science aims at understanding are not claiming that the goal of science is for scientists or laypeople to have a particular *feeling*. Rather, the sense of understanding relevant for science is that which arises when we understand a scientific theory and are able to use that theory to generate (or at least appreciate) scientific explanations (and predictions) of various phenomena. Such understanding is quite often accompanied by a feeling of understanding, but it may not always be. The key thing is actually having the relevant abilities with respect to scientific explanations and predictions.

An Underlying Problem

Thus far we have seen that the feeling of understanding is not perfectly reliable, but it might well be reliable enough to provide a good indication of when genuine understanding is present. That said, one might worry that there is a deeper problem here. Trout's criticism that understanding doesn't play an important role in science perhaps reveals something that is more troubling. If our feeling of understanding isn't perfectly reliable (and may in fact be pretty unreliable), how can we ever be sure that we truly understand some phenomena? We might be able to give scientific explanations and make accurate predictions, but can we ever be *certain* that we truly understand?

The answer to this question is "no, we cannot be certain." However, this shouldn't be surprising or troubling. All of science is uncertain. Recall from Chapter 3 that psychological certainty has to do with our level of conviction, whereas epistemic certainty has to do with our evidence. We are psychologically certain of something when we have no doubt – that is, we are absolutely convinced that we have the truth of the matter. Perhaps we can have this sort of certainty when it comes to whether we understand. The reason is that this kind of certainty is far too often unconnected to evidence or truth. There are people who, in spite of all of the evidence, are utterly convinced that the Earth is flat. Clearly, the fact that they are psychologically certain doesn't make their position any more reasonable. Thus, even though it's possible to be psychologically certain that one understands a phenomenon, that's not a good indicator of whether one truly understands, nor is it really an attitude that we should expect or desire when it comes to science. Since we can be psychologically certain even when we are badly mistaken, such certainty doesn't amount to much.

What we really want when we are concerned that the feeling of understanding leaves us uncertain is epistemic certainty. We are epistemically certain when we have the strongest possible evidence that we are correct. In other words, to be epistemically certain of something one must have evidence that is so overwhelming that it is literally *impossible* to be wrong. But, do we ever really have evidence that is this strong?

Science Is Uncertain

There are a number of factors that make science in general inherently uncertain. For one thing, science is a human endeavor, and humans are fallible. We are prone to bias, we make mistakes, and sometimes our motives are less than pure. Add to this the fact that the world we seek to understand is incredibly complex, and it naturally follows that we are always going to face uncertainties. We attempt to limit the uncertainty we face in various ways – for example, by trying to correct for biases, by employing careful methods, by utilizing precise mathematics, and so on. Nevertheless, none of these methods can completely remove all uncertainty – there's always a chance that we fail to employ these methods in a perfect manner, and on top of that it is not epistemically certain that these methods are perfectly reliable. We have already discussed above that our understanding of the nature and extent of various biases is uncertain. Additionally, there are uncertainties concerning which methods are best – there's even uncertainty about what exactly the steps in the so-called "scientific method" are supposed to be.

Physicist Carlo Rovelli nicely summed up the relation of science and epistemic certainty: "The deepest misunderstanding about science . . . is the idea that science is about certainty. Science is not about certainty . . . The very expression 'scientifically proven' is a contradiction of terms. There's nothing that is scientifically proven." Science itself is inherently uncertain. Despite this uncertainty, science produces some of our best-confirmed knowledge of the world around us. Consequently, we can have a lot of scientific knowledge despite (or perhaps, as some argue, because of) uncertainty. The lesson to draw from this is that while we may not be in a position to be epistemically certain that we have achieved genuine understanding, that doesn't mean that we haven't actually achieved such understanding, nor does it mean that scientific understanding isn't of critical importance.

Genuine Understanding and Accurate Explanations

We've seen that there is some reason to think that the phenomenological sense of understanding (feeling that one understands) is not a perfectly reliable indicator of when one actually understands. Despite this, we have also seen that such feelings might be important signposts on the road to genuine understanding. Our starting concern in this chapter was simply whether we are good judges of explanatory quality. We have seen that there is no reason to think that we are unable to spot genuine understanding, and by extension good explanations.

There is, however, one additional issue that should be addressed here. Is truth important for understanding or scientific explanations? Yes, truth is extremely important for both. While a false explanation might be such that *if* it were true it would provide understanding, genuine understanding requires accurate explanations. To borrow an example from Elgin, consider a person who understands astrology. In terms of the framework we have adopted from de Regt, this person understands the theoretical foundations of astrology. Now let's think about their attempt to explain various phenomena. For instance, imagine that the person has read their morning horoscope and consulted the alignment of the stars and uses this information to explain why they have been having a tough time getting along with their colleagues. Let's add to the case that they feel that they understand the phenomena in question (the friction with their colleagues). In fact, we can go so far as to say that they are *psychologically* certain that the explanation for their troubles lies in various facts about their horoscope and the way the stars and planets are positioned. Do they understand why they are having trouble getting along with their colleagues? No, of course not. Astrology is false, and the explanations it affords this person are inaccurate. Hence, despite the person's understanding of astrology and their strong conviction that they understand why they are having difficulty getting along with their colleagues, this person doesn't genuinely understand what is going on. Astrology fails to provide accurate explanations of the phenomena in question, but instead erroneously attempts to explain human behavior by appealing to the positions of the stars and planets. Genuine understanding requires accurate explanations. Judgments of explanatory quality (which sometimes, but not always, involve appreciating feelings of understanding) are our guides to which explanations are best, and so most likely to be accurate. This is particularly so within the context of the scientific community, where theories and hypotheses are tested and judged, steps are taken to control for biases, and ultimately consensus is reached about whether genuine understanding has been achieved. Thus, although felt understanding is not a perfect guide to explanatory quality, it shouldn't be dismissed out of hand either.

7 False Theories, But Accurate Explanations?

Theories, Models, and Explanations

In the previous chapter we discussed the importance of accurate explanations. Without explanations that are in fact accurate we cannot have genuine understanding. In this chapter we will explore whether false scientific theories can be used to generate accurate scientific explanations. Before jumping into this, let's first briefly recall the relationship between scientific theories and scientific explanations. Scientific theories consist of laws, models, and principles. Together these components of scientific theories offer broad generalizations about the nature of the world.

One thing that immediately stands out when describing models (scientific or otherwise), as we noted in Chapter 4, is that by their very nature they make simplifying assumptions about the systems that they represent. Without making such assumptions there would be no model at all – we would simply be looking at the system itself. That said, there are different sorts of simplifying assumptions that a model may make. As we mentioned, scientific models may, like a subway map, simply leave out information. In this sense many scientific models abstract away from various details of the systems they represent. One way of doing this is by approximating values for various components of the system. For example, if one is modeling a particular population of organisms, one might round the number of organisms to the nearest interval of 10 instead of working with the exact number. Another way some models simplify is by treating various aspects of the target system as having properties that are known to be false. This occurs, for instance, when developing biological models that treat a given population as infinite. Obviously, there are no infinitely large populations of living organisms, and we know this. However, at times treating a population as if it were infinite is helpful for isolating other aspects of the population that one wants to study. Such models are *idealizations*, and they are quite common in science.

One might wonder, why use idealizations that are not entirely precise and are known to be inaccurate in various ways? The short answer is that idealizations make things simpler and easier to work with. The longer answer is that, as we have discussed in earlier chapters, we face two very important challenges: the phenomena we seek to explain and make predictions about are extremely complex and humans are limited in terms of cognitive abilities, time, and resources. This combination of complexity and limitations makes idealizations a necessity for doing science.

Getting Accurate Scientific Explanations from Idealizations

Idealizations are a fixture of science. However, we saw in the previous chapter that what we really seek in science is accurate scientific explanations, because these are required for genuine understanding. These two facts may at first seem to be in tension. Fortunately, it is possible to use idealized models to generate accurate scientific explanations. To see why, let's first recall what an explanation is in general. As we saw in Chapter 2, an explanation consists of information about dependence relations obtaining between the phenomenon being explained and other things (earlier conditions, natural laws, and so on). Such information is what allows us to answer "how" and "why" questions about the phenomena we seek to explain and understand. Knowing information about dependence relations doesn't just allow us to answer "how" and "why" questions, it also allows us to answer important questions about how things might have been different or would be different if various changes occurred. Hence, an accurate explanation is one that provides accurate information about dependence relations, and so provides accurate answers to "why," "how," and "what-if-things-had-been-different" questions. Idealized models help us come to possess accurate information of this sort.

Recall that a model is an idealization when it either abstracts away from various details (it leaves things out or perhaps employs approximations rather than exact figures) or assumes values that we know to be false. In both cases what a successful idealization is doing is providing us with information about dependence relations. Let's consider a concrete example that is common in population genetics. Often, we want to represent how natural selection works by determining what would happen if we were to look at just the fitness (a concept that relates to the propensity to survive and reproduce) of a particular trait and the frequency of that trait in a given population. In models of population genetics, it is common to idealize by assuming that the population in question is infinitely large. As we noted above, this is not the case for any actual population of organisms. Nevertheless, by making this idealization we are able to home in on the dependence relations obtaining between the fitness and the initial frequency of a trait in a population on the one hand, and the resulting distribution of that trait in the population on the other. By idealizing evolutionary forces in various ways in a particular model, we are able to come to have accurate information about dependence relations that actually exist in the system being modeled, such as the fitness of a particular trait, which might be obscured by other factors – for example, genetic drift-in samples of smaller populations. The general idea applies to idealizations in all areas of science – by idealizing various factors we are able to isolate particular features of a system and determine the dependence relations that obtain between those features and other features of the system of interest. Essentially, idealized models offer a way of doing what we seek to do in *any* precise scientific experiment – that is, control for confounding variables. Idealization offers a way of exploring dependence relations obtaining between dependent variables and particular outcomes.

One final point is worth briefly discussing before moving on. When we employ an idealization, we *know* that a particular parameter has been approximated or falsified in a particular way. Thus, idealizations don't lead us to mistakenly think that there aren't dependence relations where there really are. Instead, idealizations let us discover dependence relations or aspects of those relations that might be obscured otherwise.

Does This Really Lead to Understanding?

Idealizations are necessary, and they provide us with accurate information about dependence relations. Nonetheless, idealized models intentionally leave out information or work with false information. As a result, any scientific explanation generated by an idealized model must be incomplete in various ways. It will fail to provide a complete and precise picture of what is going on. But, this is unsurprising and no cause for concern. Given our limitations and the complexity of the world, *any* scientific explanation will leave out some information and so be incomplete. We simply aren't in a position to have the complete story (when this is understood to include *all* of the information) about any actually occurring phenomenon. For instance, the actual *complete* story of a particular event will include all the details about each and every causal influence on the event. And, it will also have to include all the details for each of the causal influences on every causal influence on the event, and so on. We don't, and can never, have a complete story in this sense.

Given this, can idealizations really lead us to genuine understanding of phenomena? As we have already noticed, science always leaves us with uncertainties. Consequently, the best way of thinking about the success of science is as providing us with deeper understanding of the world as opposed to something like giving us knowledge with certainty. Science has continued to be successful because it has continually generated deeper and deeper scientific understanding. The history of science shows scientific progress despite science *never* providing us with epistemic certainty. Hence, we can see that science advances even though the theories and models it produces, as well as the explanations and predictions that are drawn from those theories and models, are uncertain. Scientific understanding requires accurate explanations, but those explanations don't have to be complete. Furthermore, scientific understanding is something that comes in degrees. Hence, one can have more or less scientific understanding of a particular phenomenon. Our scientific understanding of a particular phenomenon can improve even if the theories and models which facilitate that understanding are false.

Philosopher Catherine Elgin has provided a helpful illustration of this fact by looking at advancements in the history of astronomy. The Copernican (heliocentric) model of the solar system constituted a scientific advance because it yielded deeper understanding than the earlier Ptolemaic (geocentric) model of the solar system. Importantly, this was achieved despite the fact that the Copernican model included false information about important features of the solar system. For instance, the Copernican model included the idea that Earth's orbit around the sun is circular. Johannes Kepler improved the Copernican model by replacing the idea that the Earth's orbit is circular with the idea that it is elliptical. This was an advance in scientific understanding even though, strictly speaking, this idea is false as well! When Newton used his theory of gravitation to hit upon the truth that the Earth's orbit isn't exactly elliptical, this marked another advance in scientific understanding despite the fact that general relativity has replaced Newton's theory of gravitation. In each of these cases we see science advancing and scientific understanding improving in spite of the fact that each of the models and theories along the way were false (or, more carefully, were false about certain key claims). Although newer models and theories work better and provide more understanding than the older models and theories they replace, this doesn't mean that the older ones didn't work at all or provided no understanding.

It is worth emphasizing the general lesson we learn from these advances in astronomy: science and our understanding of the world can advance by way of models and theories that are, strictly speaking, false. This is so because some false theories and models are closer to being correct than others. Sometimes two false theories or models of the same phenomena differ in the depth of understanding they yield. Consider a simple example. Imagine two children with mistaken views of evolution. The first child believes that humans descended from chimpanzees. The child is mistaken because humans didn't descend from chimpanzees, but rather humans and chimpanzees share a common ancestor. The second child believes that humans descended from gorillas. This child is also mistaken. But, when compared to the first child, this child is further from the truth because humans have a more recent common ancestor with chimpanzees than they do with gorillas. As a result, when we consider the two children's understanding of evolution it is clear that the first has a deeper understanding than the second. And, they both have a better understanding than a third child who believes that humans sprang into existence out of rocks.

Therefore, scientific understanding can advance even by way of false theories and models, which can generate scientific understanding and contribute to the advancement of science. Similarly, the approximations or intentional falsehoods in idealizations do not render them incapable of producing genuine scientific understanding. Science can and does readily make use of models that are simplifications or distortions in order to better understand features of various phenomena. Idealizations are not an obstacle to generating scientific understanding; they are among our most common methods of attaining such understanding.

Uncertainty Remains

The world that science seeks to explain is exceedingly complex and we are limited in a number of important ways. We don't have all of the evidence, we can't observe the past or the future, we can't observe the microphysical, we can't even observe all that is happening at the macro level in our universe, and on top of all of this we are limited in our ability to process the evidence that we do have. However, none of this means that we cannot have good, accurate scientific explanations or make good, accurate predictions. Further, none of this means that we can't genuinely understand phenomena or possess scientific knowledge. What this does mean is that we must make use of models in order to get a handle on the phenomena that we desire to understand. All models make use of idealizations and simplifying assumptions. Such assumptions will always leave us with some uncertainties. This isn't a flaw of idealized models though; it is simply the nature of science. As physicist Carlo Rovelli has said, "the core of science is not certainty, it's continual uncertainty." It is this uncertainty that helps science advance. Idealizations have played, and will always play, a critical role in the advancement of science.

8 From Explanation to Knowledge

Inference to the Best Explanation

It is widely held that science is a (if not *the*) primary source of our knowledge of the world around us. Further, most accept that scientific knowledge is the best confirmed and well-supported kind of knowledge that we have of the world. But, how do scientific explanations lead to scientific knowledge? The short answer is that they do so via an inference method known as "inference to the best explanation" (IBE), sometimes called "abduction." Before we get into the details of IBE, let's take a quick look at an obvious way that scientific explanations give us scientific knowledge.

Clearly, when we know that a particular scientific explanation is accurate, we can use that explanation to give us descriptive knowledge. In other words, we can come to know what happened in a particular situation when we possess an accurate scientific explanation of the event. For instance, if we know that a solid object is held together because of attractive forces between its molecules, we can use this knowledge to explain why a sugar cube dissolved in a cup of hot tea and thereby come to know why/how the cube dissolved. Since we know that the sugar cube is solid because of attractive forces, we can use that knowledge (along with other bits of knowledge) to recognize that the binding of the sugar cube's molecules with some of the tea's molecules depends upon the differing polarizations of the sugar molecules and water molecules, and so on.

We can also use knowledge of an accurate scientific explanation to make predictions. While we may not know whether the prediction will actually be borne out or not, we can know (because of our knowledge of the accurate scientific explanation) how likely it is that the predicted event will occur. Take the simple sugar cube example again: by using our knowledge of the accurate scientific explanation of the sugar cube's dissolving, we can accurately predict how likely it is that some other sugar cube will dissolve when dropped into another cup of hot tea. Such knowledge (of what has happened and what might happen) is undoubtedly very important.

Another way that scientific explanations provide scientific knowledge is when our evaluations of them allow us to infer that the best explanations are correct. Knowledge of which scientific explanations are correct in turn allows us to determine which scientific models and scientific theories are correct, which is a major part of what we typically mean by "scientific knowledge." The general idea of IBE is that the theoretical virtues discussed in Chapter 5 are reliable guides to the truth. A bit more precisely, IBE is the idea that the explanation that would, *if true*, provide the deepest understanding of some phenomenon, is the correct explanation of that phenomenon. More precisely still, IBE is often characterized in this way (though, as we will see below, this common formulation needs some adjusting in order to be acceptable):

IBE

1. There is a set of data that is in need of explanation.
2. Hypothesis H would, if true, explain the relevant data.
3. There is no available rival hypothesis that explains the relevant data as well as H does.
4. Therefore, H is (probably) true.

Before we continue, let's get clear on a couple of points about IBE. First, the fact that a hypothesis would, if true, explain some data does not entail that the hypothesis is the *actual* explanation of the data. Rather, this means that the hypothesis is a potential explanation of the data. After all, if H would explain the data, in order for some other hypothesis to be a rival of H that hypothesis would also have to offer a potential explanation of at least some of the data as well. However, it can't be that both H and rivals to H (i.e., hypotheses that potentially explain the same data but are inconsistent with H) are actual explanations of the data. For example, it can't be that life on Earth originated with RNA preceding proteins and that proteins preceded RNA in the origin of life. At most one of these hypotheses provides the actual explanation of the origin of life on Earth (of course, it may be that neither is correct). In any case, the actual explanation of some data is true, but it can't be that multiple rival hypotheses are all true. Instead, at most one of a set of rival hypotheses can be true. As a result, the idea in IBE is that H, and its rivals, are potential explanations of the data – they would each, if true, explain at least some of the data. Thus, IBE is saying that it is legitimate to infer that the hypothesis that is the best potential explanation of the relevant data is the actual explanation of the data.

Second, we might wonder where the "best" comes into inference to the best explanation, since the term doesn't appear in premises 1–3 of IBE at all. Premises 2 and 3 are where this idea comes into the picture. Premise 2 says that H explains the relevant data, and premise 3 says that no available rival hypothesis explains the data as well as H does. Since H explains the data and no rival explains it as well as H, H offers the best explanation of the available hypotheses. H is the best explanation of the data because it offers an explanation of the relevant data, and it is more theoretically virtuous (it possesses to a greater degree virtues such as explanatory power, predictive power, conservatism, simplicity, and beauty) than any of the rival hypotheses.

Detectives, Mechanics, and the Rest of Us

Inference to the best explanation occurs so often that it is plausible that we usually don't even recognize that we are doing it. You come home and there are groceries in your kitchen. You infer that your spouse has been to the market. Why? This hypothesis (that your spouse has been to the market) best explains the data: the bags of groceries in your kitchen, the fact that your spouse said that they might go to the market today or tomorrow, the fact that your spouse is the only one who is at home when you arrive, and so on. As we saw in Chapter 5, this sort of reasoning is not only common, but it occurs even in very young children. Some studies suggest that even infants might engage in a rudimentary form of this kind of reasoning! In order to get a firmer grip on IBE, let us consider a few additional commonplace examples that exemplify reasoning that fits the schema:

i. Sherlock Holmes is investigating a crime. Dr. Watson has told him that he saw Moriarty leaving the crime scene from the north side of the building right after the crime was committed, and he didn't see anyone else enter or leave the building. Holmes knows that Moriarty stood to gain from the crime being committed. A thorough search of the scene has revealed only one set of footprints, which are the same size as Moriarty's. Holmes had been watching the building from the south side and no one came into or out of the building immediately before or after the crime. Furthermore, Holmes knows that there are only two entrances to the building, the one on the south that he was watching and the one on the north that Watson was watching. Holmes infers that Moriarty committed the crime.

ii. Michelle the mechanic is examining a particular car that continually has a dead battery. She has checked and all of the wiring is in good shape. Michelle has also tried putting several different new batteries in the car with the same result – after a short period of time the battery is completely dead. She infers that the car has a faulty alternator.

iii. You and a colleague have been given the same assignment at work. The assignment is relatively simple, but it is time consuming. You and your colleague work at equally fast paces, and the assignment took you all of last week in order to meet the Monday deadline. You find out that your colleague failed to complete the assignment by the deadline. You infer that they didn't work on the assignment for as many hours as you did last week.

In (i), (ii), and (iii) the inference drawn is perfectly reasonable. In each case, the reasoning fits the IBE schema. For instance, here's the Holmes inference (the others similarly fit the schema):

Holmes' IBE

1. (a) Watson saw Moriarty leaving the scene right after the crime; (b) no one else entered or left the north side of the building; (c) no one entered or left the south side of the building; (d) Moriarty had a motive for committing the crime; and (e) footprints matching Moriarty's were found at the crime scene.
2. Moriarty's committing the crime explains (a)–(e).
3. There is no available rival hypothesis that explains (a)–(e) as well as that Moriarty committed the crime.
4. Therefore, Moriarty (probably) committed the crime.

Of course, there could be rival hypotheses concerning the crime. Perhaps a lookalike is trying to frame Moriarty. Similarly, it could be that the other mechanics at the shop are playing a trick on Michelle. And, it could be that your colleague worked the same amount of hours as you but somehow they managed to get less done on this one occasion, even though the two of you have always worked at the same speed in the past. Each of these hypotheses offers a *possible* explanation of the relevant data in the particular situations discussed. But it seems that the hypotheses that were inferred to be true in (i), (ii), and (iii) are better than these rivals, given how the cases are described. In light of this, although these other hypotheses (the Moriarty lookalike, prankster mechanics, and your colleague's limited productivity) are possible explanations of the relevant data, the reasonable thing in each of these cases is to infer the hypothesis that was actually inferred. Holmes, Michelle, or you could be wrong, but this doesn't mean that it wasn't reasonable to infer as each of you did. In each of these cases, the inference fits the IBE schema, and it is clear that the inference seems perfectly reasonable. We could go on describing everyday instances of IBE indefinitely.

Inference to the Best Explanation in Science

In many ways the practice of science is quite different than our ordinary lives. Scientists often make use of controlled experiments in labs, they have high-tech equipment for making precise measurements, and so on. We didn't see anything of this sort in the cases of IBE described above. And, IBE makes no explicit mention of such practices. In light of this, do we still have good reason to think that IBE is used in science?

The answer is "yes." IBE is very common in science. We have already noted, but it bears repeating, that Charles Darwin's argument for natural selection was an IBE. At the time Darwin was writing, the generally accepted explanation of the diversity of life was that there was a divine act of special creation of each species. Darwin's theory, however, offered the best explanation of the data from anatomy, development, biogeography, and fossils, among other things. As a result, the consensus among biologists (at least after a period of time) has been that Darwin's theory is correct because there is no rival that explains the relevant data nearly as well as it does. When a particular scientific theory, evolutionary theory, say, generates hypotheses that offer really good explanations of large sets of data, such as homologous structures in anatomy and the fossil record due to common descent, and those hypotheses better explain the data than the hypotheses generated by rival theories such as homologous structures being the implantation of an archetype, the rational thing is to infer that that scientific theory is true.

IBE has also been used to facilitate major discoveries. For example, as we discussed in Chapter 1, toward the beginning of the nineteenth century, when there were only seven known planets in our solar system, scientists discovered that the orbit of Uranus departed from what it should be, given the truth of Newton's theory of universal gravitation and the assumption that there were no additional planets (beyond the known seven) in the solar system. One hypothesis for this deviation was simply that Newton's theory is false. However, this theory had had great empirical successes for (then) more than two centuries. So, this hypothesis seemed to run afoul of the theoretical virtues of conservatism (because it required going against many well-supported findings) and explanatory power (because it provided no alternative explanation of the empirical successes of Newton's theory). A different hypothesis was that the explanation for the deviation in Uranus's orbit was the presence of an eighth, as yet undiscovered planet in the solar system. When the explanations offered by these two hypotheses were compared, the second was much more theoretically virtuous. As a result, the best explanation, that there was an eighth planet, was inferred to be true. Not long after this inference had been made the planet Neptune was discovered.

Another famous discovery as a result of IBE was the discovery of the double-helix structure of DNA. Positing a double-helix structure best accounted for a variety of evidence, from the evidence from X-ray diffraction for a helical structure to the observation that the bases A and T were found in equal amounts in a DNA molecule, as were G and C. There are many more examples of IBE leading to major scientific discoveries – several of which we discussed briefly in earlier chapters (particularly Chapter 5) such as the grounds for replacing the Ptolemaic (geocentric) model of the solar system with the Copernican (heliocentric model), the discovery of the electron, and so on. The list of important IBEs in science is long indeed. In fact, some, such as

J. D. Trout, go so far as to argue that the success of modern science as a whole rests upon its use of IBE.

Of course, our scientific knowledge is tentative. This is a feature of all knowledge in science – we may revise our scientific theories or accept new scientific theories in light of new data or the development of new rival theories. After all, even the best of science is just that – *science*; it isn't dogma that we believe regardless of new evidence. Nevertheless, the fact that scientific knowledge always remains open to future revision doesn't undermine the fact that it is genuine *knowledge*. Remaining open to new evidence that might challenge what you already know instead of simply dismissing such evidence out of hand isn't a weakness of scientific knowledge, but rather it's a virtue of those who come to possess scientific knowledge.

Can We Be Sure That the Best Is Good Enough?

A number of objections have been raised against IBE. The aim of such objections is to show that inferring to the best explanation is unreasonable in general. If IBE really does license unreasonable inferences, then we have a problem when it comes to our everyday inferences as well as our understanding of scientific knowledge (and our claims to possess such knowledge is threatened). Let's take a close look at some of the most prominent challenges to IBE.

Let's begin with philosopher Bas van Fraassen's "Best of a Bad Lot" objection:

> To believe is *at least* to consider more likely to be true, than not. So to believe the best explanation requires more than an evaluation of the given hypothesis. It requires a step beyond the comparative judgment that this hypothesis is better than its actual rivals . . . For me to take it that the best of set *X* will be more likely to be true than not, requires a prior belief that the truth is already more likely to be found in *X*, than not.

We might understand van Fraassen's challenge in two distinct ways. The first way of understanding it is simply that it is pressing the "why think the world is simple/beautiful?" challenge that we discussed in Chapter 5. Since we have already explored this objection and found it wanting, let's focus our attention here on the other way that we might understand van Fraassen's challenge: that when we infer the truth of the best explanation, we may have considered only bad explanations. "Best" is a comparative term. So, saying that H provides the "best" explanation only means that it provides an explanation that is better than the available alternatives. The best available might still be bad. Put another way, the concern is that it may be that H provides the best explanation among those offered by the hypotheses we have considered, but we need good reason to think that the true hypothesis is likely to be among those considered before we are warranted in inferring that H is true. It is important to consider the Best of a Bad Lot for at least two reasons: (1) it is often raised as a problem for IBE and (2) though it ultimately fails to impugn IBE, it does make it clear that the general schema of IBE is lacking an important qualification.

A straightforward response to the Best of a Bad Lot is to add a restriction that simply being the best available explanation of the relevant data isn't enough to warrant inference. Instead, IBE should be understood to say that the best explanation must be a sufficiently good explanation in its own right before it is reasonable to infer that it's true. As a number of supporters of IBE have emphasized, IBE should really be understood as saying that the best explanation of some data is true when that explanation is *good enough*.

At this point, two questions immediately present themselves. The first is simply how could van Fraassen, or anyone else, ever think the Best of a Bad Lot was a genuine problem for IBE? After all, isn't it obvious that in order for inference to be warranted the best explanation has to be a good explanation? The second question concerns the idea of an explanation being "good enough." What exactly is required for an explanation to meet this standard?

The first of these questions has a fairly straightforward answer. Van Fraassen, and others, challenged IBE along these lines because discussions of IBE had been a bit sloppy prior to his raising this objection. Even supporters of IBE often failed to present the inference form as carefully as they should have. As a result, the critical qualification that the best explanation be "good enough" was often (and sometimes still is) left out of formulations of IBE. A lesson to be learned from the Best of a Bad Lot is that care is needed when thinking about and presenting IBE. With this lesson in mind, let's formulate IBE in the way that it should be – that is, a way that takes account of the needed "good enough" qualification:

IBE

1. There is a set of data that is in need of explanation.
2. Hypothesis H would, if true, explain the relevant data *and* H is a good enough explanation of the relevant data.
3. There is no available rival hypothesis that explains the relevant data as well as H does.
4. Therefore, H is (probably) true.

This revised formulation of IBE suggests that we might think of it as involving two stages. In stage one, we formulate a number of initially plausible hypotheses that offer potential explanations of the relevant data. At this stage we select hypotheses that fit with the information that we have and dismiss wildly outlandish hypotheses, such as philosophers' worries that perhaps the world is just a dream or simply a computer simulation. Then, in stage two, the hypotheses that are plausible in light of the background information that we have are assessed in light of their theoretical virtues.

When it comes to our second question – how good an explanation must be in order to be "good enough" – things aren't so easy. It seems clear that the sorts of theoretical virtues discussed in Chapter 5 will certainly play a critical role here. After all, such virtues are what determine the quality of any given explanation. That said, it is a very difficult matter to delineate when an explanation is good enough to be inferred (assuming that it's also the best) and when an explanation isn't good enough to be inferred (despite being the best). In order to appreciate one of the suggestions for responding to this difficulty, let's consider a distinct challenge for IBE that is similar to the Best of a Bad Lot in demonstrating the need for a "good enough" qualification.

Another challenge for IBE, called the "Disjunction Objection," makes plain the fact that simply being the best explanation of the relevant data is not enough to warrant inferring that a hypothesis is true. Although this challenge for IBE exposes the same sort of weakness in our original formulation of IBE as the Best of a Bad Lot (i.e., it demonstrates the need for the "good enough" qualification), the two challenges are different in important ways. As we have seen, the Best of a Bad Lot revolves around the concern that the true explanation of the relevant data may not be among the set of available hypotheses. Hence, the hypothesis that is the best of the lot that we have considered may fail to be the actual explanation of the data we are examining. The Disjunction Objection is different because the concern that it raises is an issue even if it is *epistemically certain* that the true hypothesis is among those that we have considered.

In order to get a firm grasp of this challenge, let's consider an example concerning the origin of life on Earth. For the purpose of depicting this challenge, let's assume that there are only three possible hypotheses: life on Earth began with RNA (RNA world); proteins proceeded RNA in the origin of life on Earth (Protein world); and RNA and protein evolved together (RNA & Protein world). So, this is a situation where van Fraassen's Best of a Bad Lot concern cannot arise because we are assuming that at least (and at most) one of these hypotheses is true. This doesn't necessarily guarantee that we should infer that the best explanation of the three hypotheses is true though. After all, it could be that RNA world, the best explanation, say, is more likely than Protein world, and it is more likely than RNA & Protein world; but RNA world isn't more likely than the disjunction: Protein world or RNA & Protein world. To see this, let's assume that the probability that RNA world is true is 0.4, the probability that Protein world is true is 0.3, and the probability that RNA & Protein world is true is 0.3. In such a case, it may well be that RNA world is a better explanation than either Protein world or RNA & Protein world, but it's still unlikely to be true. After all, it is 0.6 probable that the correct explanation is Protein world *or* RNA & Protein world. In other words, it's 60 percent likely that RNA world is false. Thus, RNA world, although the best explanation, is more likely than not to be false. Plausibly, IBE doesn't license inferring RNA world in this situation because although it's the best explanation, it's not "good enough."

Considering the Disjunction Objection helps illuminate one way in which we might understand the "good enough" requirement for IBE. The fact that a particular hypothesis is the best explanation of the relevant data warrants inferring that the hypothesis is true *only if* the theoretical virtues of that hypothesis make its truth sufficiently probable. How probable a hypothesis needs to be before it should be inferred will likely depend upon the circumstances. Another answer for when a hypothesis is "good enough" appeals to an additional IBE. Philosopher Finnur Dellsén has argued that inferring that the best explanation, H, say, is true is warranted only when the best explanation of our inability to think of a better explanation is that there is no better explanation to be had. Essentially, Dellsén's idea is that the explanation provided by a particular hypothesis is good enough to warrant inference when, taking account of our research and work in coming up with rival hypotheses, it is reasonable for us to believe that no better hypothesis could be generated.

There is still work to be done in determining exactly how we should think of the "good enough" qualification, but these are two promising (and nonexclusive) routes we might go. Furthermore, it may be difficult to determine where exactly the line between good enough and not good enough should be drawn. Likely, what's good enough will depend upon the particular circumstances in which we find ourselves. It may be that the explanation offered by a hypothesis would be good enough to warrant inferring the truth of the hypothesis in one context but not in another. Perhaps the practical implications of accepting a particular hypothesis as true affect whether it is "good enough" to infer as well. And, there will almost certainly be tough cases where it is difficult to be sure whether the explanation offered by a hypothesis is good enough to warrant inferring its truth. It is likely that in such cases there will be disagreement about the status of the hypothesis. This shouldn't surprise us. Science, like all of life, is hard sometimes. Fortunately, if we hit one of these tough cases, we can always gather more data. This may allow us to better determine the quality of an explanation. After all, the quality of explanation provided by a hypothesis is apt to change (for better or worse) with the accumulation of more data. And, of course, there will be (and are) many cases in science where it is clear that the best explanation is good enough to warrant inference and many cases where it is clear that the best isn't good enough.

What If There Are Lots of Good Explanations?

We began the previous section by considering a worry for IBE that arises because we may not be sure that the potential explanations we have available are very good. Another sort of concern that is sometimes raised comes at IBE from the opposite direction: What if there are lots of good explanations on offer? Let's assume, at least for the sake of illustration, that the RNA world hypothesis we discussed in the previous section is the best explanation of the origin of life. Despite being the best, it doesn't seem that we are warranted in inferring that the RNA world hypothesis is true. The fact that there are other plausible rivals (the Protein world hypothesis and the RNA & Protein world hypothesis, for instance), even if they aren't as good as the RNA world hypothesis, seem to make it so that we shouldn't infer that the hypothesis that provides the best explanation is true. In a similar vein, recall the Alvarez meteorite-impact hypothesis that explains the K–Pg (Cretaceous–Paleogene) extinction. Assuming that the Alvarez hypothesis provides the best explanation of the K–Pg extinction, this isn't sufficient to warrant accepting this hypothesis as true because there is at least one plausible rival hypothesis: volcanic activity led to the K–Pg extinction, or at least contributed to it along with the meteorite impact.

Instances where there are multiple plausible rivals again suggest that simply being the best explanation isn't sufficient to warrant inferring the truth of a particular hypothesis. That said, the "good enough" qualification seems to address this problem as well. For instance, think of the RNA world hypothesis; given the plausibility of its rivals, this hypothesis doesn't seem to meet either of the standards for "good enough." It is likely that its probability, while higher than either of its chief rivals, still falls short of being high enough to warrant accepting the hypothesis as true. Further, the fact that there are plausible rivals makes it likely that we aren't in a position to reasonably conclude that no better hypothesis could be formulated in the future. Consequently, IBE doesn't license inferring that the RNA world hypothesis (or either of its rivals) is correct. We need more data to establish that this hypothesis (or some other) really is *good enough* to warrant inferring its truth.

Aren't There Always Other Explanations That We Haven't Thought of Yet?

Let's consider one additional challenge to IBE that seeks to undermine its legitimacy. This challenge concerns alternative hypotheses that we haven't considered (often because the hypotheses haven't been thought of yet). Here is how van Fraassen has expressed this challenge:

> I believe, and so do you, that there are many theories, perhaps never yet formulated but in accordance with all evidence so far, which explain at least as well as the best we have now. Since these theories can disagree in so many ways about statements that go beyond our evidence to date, it is clear that most of them by far must be false. I know nothing about our best explanation, relevant to its truth-value, except that it belongs to this class. So I must treat it as a random member of this class, most of which is false. Hence it must seem very improbable to me that it is true.

The thrust of van Fraassen's concern is centered on the worry that there are innumerable hypotheses that we haven't thought of yet. In fact, he went so far as to argue that many of these undiscovered hypotheses offer explanations that are as good as, or better than, those provided by our current best scientific hypotheses. Thus, van Fraassen has argued that, given all of these really good undiscovered hypotheses, we can't be sure that our current "best" hypothesis really provides the best explanation of the data. And, he thinks that we certainly cannot reasonably think that a particular hypothesis is true because it offers the best explanation among the hypotheses we have considered so far.

The concern that van Fraassen has raised does have some truth to it. After all, even a cursory examination of the history of science reveals that the best theories and hypotheses of a given period are often later replaced by even better theories and hypotheses. Given this and the fact that we should always be open to revising in light of new evidence, it is certainly possible that any of our current best scientific theories and hypotheses *might* one day be replaced. In spite of this, van Fraassen is mistaken in asserting that this should give us cause to question the legitimacy of IBE. First of all, it is far from clear that our current best scientific hypotheses really should be treated as just members of a large set of predominantly false hypotheses. Science has advanced not only in terms of coming up with better theories and hypotheses, but also in terms of having better equipment and stricter standards for testing hypotheses. Science is cumulative. Consequently, we possess more scientific knowledge now than was possessed in the past. This additional scientific knowledge that we already possess forms a background against which we formulate new scientific theories and hypotheses. The richer our store of background knowledge, the more likely the hypotheses that offer us the best explanations are to be true.

With the above being said, even if we grant that there are other equally good explanations that we haven't thought of yet (an assumption that is far from clearly true!), it doesn't follow that IBE is untrustworthy. At most the truth of this assumption would entail something that we should already recognize: IBE is fallible. Sometimes the best available explanation of the data turns out to not be true – something that additional data and additional hypothesizing reveals. But all human reasoning is fallible in this way. We sometimes make mistakes. Do we think that because we sometimes fall prey to optical illusions that we can't get knowledge by way of observation? No, of course not. Similarly, IBE isn't perfect – that is, sometimes the inferred hypothesis will turn out to be false, but this doesn't mean that IBE isn't a legitimate form of inference. Sometimes there are car wrecks and sometimes airplanes crash, but this doesn't mean that driving and flying aren't good ways of traveling. When we walk we sometimes trip, but that doesn't mean that we shouldn't walk! When all is said and done, what matters isn't that IBE *sometimes* leads us to wrong answers, but rather that IBE quite often leads us to the right ones.

Can We Really Know?

The reader is sure to have noticed that the conclusion of IBE is "H is (probably) true." It's important to include the parenthetical "probably" because supporters of IBE disagree on whether the knowledge we get from IBE is outright or only probabilistic. Some think that when a particular hypothesis, H, is the best explanation of the relevant facts and a good enough explanation, then we should infer that H is true; others claim that in this situation we should rather only infer that H is *probably* or *highly likely* or perhaps *X percent likely* to be true. Philosopher Alexander Bird has argued that when it comes to IBE the only time we should conclude that H is true – the only time that IBE gives us knowledge that a particular hypothesis is true – is when we are inferring to the *only explanation*. On Bird's view, a special limiting case of IBE is one which matches our above formulation *and* has the feature that there are no genuine rivals to H. What counts as a "genuine" rival will vary depending on the context. But, at a minimum, a genuine rival hypothesis must conflict with H and have at least some plausibility. Hence, while Bird thinks that we may not be able to definitively show that philosophers' worries, such as being in a computer simulation or that the universe recently popped into existence looking like it's roughly 14 billion years old, are false, these "hypotheses" don't matter because they don't really count as genuine rivals since they are so implausible. Nevertheless, when it comes to cases where we have actual scientific hypotheses that are rivals, we can only come to know that one of them is true when it is the best explanation, good enough, *and* we have been able to definitively show that the others are false. Any time there are additional hypotheses that could explain the data, at most we can infer that the best explanation is *probably true*.

There is no need to settle the issue of whether Bird's take on things is correct here. Instead, for current purposes, it is enough to recognize that what he says is plausible and fits well with scientific practice in many cases. There's a reason why a lot of scientific theories are said to be X percent likely to be true rather than flat-out asserted to be true. Further, Bird's position is consistent with IBE being a (even *the* primary) vehicle for gaining scientific knowledge. In some cases, our scientific knowledge is full-stop – we know that H is true; in other cases, our scientific knowledge is probabilistic – we know that it is X percent likely that H is true. Either way we come to possess scientific knowledge by inferring to the best explanation.

Is It Really Science If It Doesn't Explain Everything?

Before closing this chapter it is worth pausing to briefly consider a lingering worry that one might have. Some people worry that a particular theory or hypothesis is unscientific or somehow defective if it doesn't explain everything that we might want to know about. For instance, this sort of worry sometimes gives rise to a misguided objection to evolution. The gist of this objection is that evolutionary theory fails to explain various things (for instance, how exactly the common ancestors of humans and apes looked and how life originated in the first place), and therefore evolutionary theory is inadequate as a scientific theory, or even false.

Considering what has been said about IBE, it is easy to see where this sort of worry goes wrong. A scientific hypothesis or theory may, as evolution does, best explain the relevant facts, explain those facts extremely well, and easily exceed the standards for being good enough without explaining everything that we might want explained. There is nothing in IBE that says that a hypothesis must explain everything in order to be legitimately inferred to be true. No scientific theory or hypothesis explains everything! To appreciate just how excessively demanding this requirement is, think about an everyday situation. You observe your friend eating a salad. Let's say that the best explanation you have for this is that your friend is hungry. That's a very plausible explanation of your friend's eating. Does this explain everything though? Of course not. Why does your friend pick a salad rather than something else to eat? Why does your friend eat when hungry rather than waiting until later? Why is your friend eating a salad here rather than taking it somewhere else? Why hasn't your friend offered you something to eat? And so on. The hypothesis that your friend is hungry doesn't answer any of these additional questions. Does this mean that the hypothesis that your friend is hungry is inadequate as an explanation of their eating a salad? Obviously not. Similarly, the fact that evolution, or any scientific theory or hypothesis, doesn't explain everything in no way diminishes the plausibility of the theory or hypothesis. There are always more questions to answer, data to gather, hypotheses to form, and explanations to be made. It's this continual uncertainty that drives science to new and exciting discoveries.

Concluding Remarks

Summing up, in this book we have been doing some explaining – we have answered a number of "why" and "how" questions. Along the way we have cleared up a number of common misunderstandings (see the summary that follows for a list of these). Most importantly, we have explained why science seeks to explain the world – because this is how we come to have scientific understanding of the world. Additionally, we have explained "how" science explains the world – quite often it is by way of idealizing models that allow us to generate scientific hypotheses which provide explanations of phenomena. Once we have scientific hypotheses that potentially explain some phenomenon, we utilize our ability to evaluate the explanatory merits of hypotheses to determine which explanation is best. Then, we legitimately infer that the best explanation, if it is sufficiently good in its own right, is (probably) true. This is how we come to have scientific knowledge, and it is this scientific knowledge of explanations and hypotheses that yields genuine scientific understanding of phenomena. Although, as we have noted time and again throughout this book, uncertainty remains (because it always does) and there are many more questions and details that are worth exploring, we have made considerable progress in both understanding why and understanding how science explains the world.

What we have seen about how science works and the ever-present reality of uncertainty can help us navigate difficult situations such as the COVID-19 pandemic. By appreciating how science explains the world, we can understand why the expert advice on how best to respond to SARS-CoV-2 changed over time. This isn't because the pandemic is just a media-fueled hoax, as some extremists have claimed; it is because our understanding of the nature of SARS-CoV-2 and its epidemiology has increased as more data have been gathered and evaluated. As data are gathered and potential explanations compared and evaluated, we come to know more about SARS-CoV-2 and how to best protect ourselves from the virus. Through experimental trials we have learned how to better treat COVID-19, and fortunately, such trials have led to the development of effective vaccines. During the COVID-19 pandemic we have not only observed first-hand how and why science seeks to explain the world, but we have also, unfortunately, seen how misunderstandings about this process have led to poor decisions about safety protocols and misguided resistance to COVID-19 vaccinations. It is too much to hope that this book will have a dramatic impact on remedying such problems (in part because the readers of this book are unlikely to be inclined toward either ignoring the advice of scientific experts or refusing potentially life-saving vaccines), but perhaps this book can help in at least a small way. The better we understand how and why science seeks to explain the world, the better we can appreciate and use the knowledge that science provides us.

Summary of Common Misunderstandings

Common misunderstandings about scientific explanation and related concepts with responses arising from this book:

Explanation is simply a description of what has happened or tends to happen. Although explanations often include descriptions of what has happened/tends to happen, they are not mere descriptions. Rather, genuine explanations provide accurate information about why things happened or tend to happen as they did as well as how things happen in that way.

Explanation in science is different in kind than ordinary explanations. Whereas scientific explanations are much more precise and typically explain a much greater range of data than our everyday explanations, they are not of a different kind. In other words, the difference between scientific explanations and everyday explanations is not a matter of kind, but it is instead a matter of degree of precision, range of data explained, and so on.

Good (real) science can explain everything. So, if a theory or domain of science cannot explain everything, then it is either bad science or not even science at all. All science – even the best-supported scientific theories and most well-established domains of science – leaves us with uncertainties. No scientific theory ever explains everything. Given our cognitive limitations, there is always more data for us to discover and more questions for which to seek answers.

Explanations in historical science are inferior to those in experimental science. In general, the line between historical and experimental science can be difficult to draw because science often employs a bit of both. Even in cases where the distinction can be clearly made, there isn't good reason for thinking that one is inherently better or worse than the other.

Explanation and prediction are symmetrical; one just looks at what has happened and the other looks at what will happen. Although explanation and prediction are related in various ways, they are not simply two sides of the same coin. In many cases, the two come apart so that it is possible to make very accurate predictions without having a good explanation, and alternatively, it is possible to have a good explanation of something that it wouldn't have been possible to predict ahead of time.

Theoretical virtues such as simplicity and beauty are merely aesthetic. Many scientists do find theories and hypotheses that are simple and have certain features that are often described as "beautiful" to be aesthetically pleasing. However, it is a mistake to think that such virtues are only aesthetic. There is good evidence that things like simplicity and beauty have led to many profound scientific discoveries. These theoretical virtues have a track record that suggests that in many contexts they are dependable signs of the truth.

The sense of understanding – that is, the feeling that one understands – means that an explanation is correct. Unfortunately, various biases and other cognitive failures provide good reason to believe that at times we feel like we understand even though our explanations are inaccurate. Generally, our level of psychological certainty may not map onto the level of our epistemic uncertainty all that well.

Since the feeling of understanding isn't perfectly reliable, understanding is not important for science. It is true that the feeling of understanding is far from being a perfectly reliable guide to possessing accurate explanations. However, the sense of understanding that is important to science isn't this phenomenological feeling of understanding. Rather, the sort of understanding that matters for science comes from grasping dependence relations between features of phenomena. This grasp of dependence relations is the primary epistemic goal of science.

Only accurate models can offer genuine explanations. No model is a perfect representation of the phenomena or system that it represents. As a result, all models involve some level of idealization. Nevertheless, despite the inaccuracies in our models they are often used to generate accurate explanations. Furthermore, such models can provide strong evidence that the theories used to generate them are true.

Inference to the best explanation is flawed because the hypothesis that offers the best explanation may in fact be a bad explanation overall. Inference to the best explanation should actually be understood to say that it is reasonable to infer that the hypothesis which offers the best explanation of the relevant data is true only when the explanation the hypothesis offers is good enough.

The possibility that the hypothesis which provides the best explanation may have many plausible rivals shows that inference to the best explanation is an illegitimate form of inference. In cases where there are multiple plausible rival hypotheses, inference to the best explanation doesn't tend to license inferring that the hypothesis which provides the overall best explanation is true. Typically, in such cases the hypothesis which provides the best explanation will still fail to be good enough because of the plausibility of the set of its rivals.

It's always possible to come up with new hypotheses that explain what our current best hypotheses explain, so we can't know that any given hypothesis is correct. The fact that there is always the possibility that we might come up with alternative explanations for some phenomena doesn't show that we cannot know that a hypothesis which best explains that phenomena is true. Instead, this fact simply means that inference to the best explanation leaves us with uncertainty. When we infer that a particular hypothesis is true because it best explains the data, we might be mistaken. We never have absolute certainty that a scientific hypothesis is true. Despite this limitation, we can still know that a particular hypothesis is true because it provides the best explanation of the relevant data. It is just that our scientific knowledge is held tentatively – we are always ready to revise in light of new evidence.

References and Further Reading

Chapter 1

On understanding: de Regt, H. (2017). *Understanding Scientific Understanding*. Oxford: Oxford University Press.

On evidence for evolution: Kampourakis, K. (2020). *Understanding Evolution*. Cambridge: Cambridge University Press.

On the nature of the SARS-CoV-2 virus and COVID-19: Rabadan, R. (2020). *Understanding Coronavirus*. Cambridge: Cambridge University Press.

On uncertainty in science: Kampourakis, K. and McCain, K. (2019). *Uncertainty: How It Makes Science Advance*. New York: Oxford University Press.

On the aims of science: McCain, K. (2015). Explanation and the nature of scientific knowledge. *Science & Education* 24: 827–854.

On explanation in science: McCain, K. (2019). How do explanations lead to scientific knowledge? In K. McCain and K. Kampourakis (Eds.). *What Is Scientific Knowledge? An Introduction to Contemporary Epistemology of Science*. New York: Routledge, 52–65; McCain, K. (2020). What is biological knowledge? In K. Kampourakis and T. Uller (Eds.). *Philosophy of Science for Biologists*. Cambridge: Cambridge University Press, 36–54.

Chapter 2

On how-possibly explanations: Brainard, L. (2020). How to explain how-possibly. *Philosophers' Imprint* 20: 1–23.

On the close connection between scientific and everyday thinking: Einstein, A. (1936). Physics and reality. *Journal of the Franklin Institute* 221: 349–382; Gauch Jr., H. G. (2012). *Scientific Method in Brief.* Cambridge: Cambridge University Press.

On the role of seeking and giving explanations in learning: Gopnik, A., Walker, C. M., Lombrozo, T., Williams, J. J., and Rafferty, A. N. (2017). Explaining constrains causal learning in childhood. *Child Development* 88: 229–246; Liquin, E. G. and Lombrozo, T. (2020). Explanation-seeking curiosity in childhood. *Current Opinion in Behavioral Sciences* 35: 14–20.

On explanations as "because" answers to particular questions: Kampourakis, K. and Niebert, K. (2018). Explanation in biology education. In K. Kampourakis and M. Reiss (Eds.). *Teaching Biology in Schools: Global Research, Issues and Trends*. New York: Routledge, 236–248.

On the ubiquity of generating and evaluating explanations: Lombrozo, T. (2006). The structure and function of explanations. *Trends in Cognitive Science* 10: 464–470.

On explanation as a primitive concept: Poston, T. (2014). *Reason and Explanation: A Defense of Explanatory Coherentism*. New York: Palgrave Macmillan.

On equilibrium explanations as non-causal: Sober, E. (1983). Equilibrium explanation. *Philosophical Studies* 43: 201–210.

On the challenges for I-S explanations: Sober, E. (2020). A theory of contrastive causal explanation and its implications concerning the explanatoriness of deterministic and probabilistic hypotheses. *European Journal for the Philosophy of Science* 10: 34.

On what-if-things-had-been-different questions: Woodward, J. (2003). *Making Things Happen: A Theory of Causal Explanation*. New York: Oxford University Press.

On the covering law model and other important accounts of scientific explanation: Woodward, J. (2021). Scientific Explanation. *The Stanford Encyclopedia of Philosophy* (Spring 2021 Edition), Edward N. Zalta (Ed.), Available at: https://plato.stanford.edu/archives/spr2021/entries/scientific-explanation.

Chapter 3

On the role of speculation in science: Achinstein, P. (2019). *Speculation: Within and About Science*. New York: Oxford University Press.

On the variety of kinds of explanations in biology: Braillard, P.-A. and Malaterre, C. (Eds). (2015). *Explanation in Biology: An Enquiry into the Diversity of Explanatory Patterns in the Life Sciences*. Dordrecht: Springer; Brigandt, I. (2013). Explanation in biology: Reduction, pluralism, and explanatory aims. *Science & Education* 22: 69–91.

On historical and experimental explanations: Cleland, C. E. (2002). Methodological and epistemic differences between historical science and experimental science. *Philosophy of Science* 69: 447–451; Cleland, C. E. (2011). Prediction and explanation in historical natural science. *British Journal for the Philosophy of Science* 62: 551–582; Reydon, T.A.C. (forthcoming). The proper role of history in evolutionary explanations. Nous. https://doi .org/10.1111/nous.12402.

On skepticism concerning whether historical explanations are scientific: Gee, H. (2000). *In Search of Deep Time*. New York: The Free Press.

On the inescapability of uncertainty in science: Kampourakis, K. and McCain, K. (2019). *Uncertainty: How It Makes Science Advance*. New York: Oxford University Press.

On paradigm shifts and the nature of change in science: Kuhn, T. S. (1962). *The Structure of Scientific Revolutions*. Chicago, IL: University of Chicago Press.

On grounds for thinking that we have good reasons to reject philosophical skepticism without experiments: McCain, K. (2014). *Evidentialism and Epistemic Justification*. New York: Routledge.

On the general nature of knowledge: McCain, K. (2016). *The Nature of Scientific Knowledge: An Explanatory Approach*. Cham: Springer.

On evolution and misguided criticisms of it: McCain, K. and Weslake, B. (2013). Evolutionary theory and the epistemology of science. In K. Kampourakis (Ed.). *The Philosophy of Biology: A Companion for Educators*. Cham: Springer, 101–119.

On biology as historical science: Sober, E. (2000). *Philosophy of Biology*, 2nd ed. Boulder, CO: Westview Press.

On the lack of a single scientific method: Strevens, M. (2020). *The Knowledge Machine: How Irrationality Created Modern Science*. New York: W.W. Norton.

On actual-sequence and robust-process explanations: Sterelny, K. (1995). Basic minds. *Philosophical Perspectives* 9: 251–270.

Chapter 4

On likelihood of a cancer diagnosis in one's lifetime: Cancer Treatment Centers of America (n.d.) Women and cancer. Available at: www.cancercenter.com/women-and-cancer#:~:text=The%20National%20Cancer%20Institute%20estimates,disease%20more%20often%20than%20men (accessed May 10, 2021).

On the close relationship between scientific explanation and prediction: de Regt, H. (2017). *Understanding Scientific Understanding*. Oxford: Oxford University Press; Douglas, H. E. (2009). Reintroducing prediction to explanation. *Philosophy of Science* 76: 444–463.

On the challenges of trying to understand the world: Kampourakis, K. and McCain, K. (2019). *Uncertainty: How It Makes Science Advance*. New York: Oxford University Press; Potochnik, A. (2017). *Idealization and the Aims of Science*. Chicago, IL: University of Chicago Press.

Chapter 5

On simplicity and its role in science: Achinstein, P. (2019). *Speculation: Within and About Science*. New York: Oxford University Press; Martens, R. (2009). Harmony and simplicity: Aesthetic virtues and the rise of testability. *Studies in the History and Philosophy of Science* 40: 258–266.

On the possibility that infants use explanatory reasoning: Baillargeon, R., Li, J., Gertner, Y., and Wu, D. (2011). How do infants reason about physical events. In U. Goswami (Ed.). *The Wiley-Blackwell Handbook of Childhood Cognitive Development Vol. 2*. Oxford: Wiley-Blackwell, 11–48.

On the role of exemplars in evaluating explanatory quality: Bird, A. (2022). *Knowing Science*. Oxford: Oxford University Press.

On the explanatory quality of evolutionary theory: Kampourakis, K. (2020). *Understanding Evolution*. Cambridge: Cambridge University Press; McCain, K. and Weslake, B. (2013). Evolutionary theory and the epistemology of science. In K. Kampourakis (Ed.). *The Philosophy of Biology: A Companion for Educators*. Cham: Springer, 101–119.

On the ubiquity of generating and evaluating explanations: Lombrozo, T. (2006). The structure and function of explanations. *Trends in Cognitive Science* 10: 464–470; Lombrozo, T. (2016). Explanatory preferences shape learning and inference. *Trends in Cognitive Science* 20: 748–759.

On the role of beauty in science and its relation to simplicity: Glynn, I. (2010). *Elegance in Science: The Beauty of Simplicity*. Oxford: Oxford University Press; McAllister, J. W. (1996). *Beauty and Revolution in Science*. Ithaca, NY: Cornell University Press; Weinberg, S. (1994). *Dreams of a Final Theory: The Scientist's Search for the Ultimate Laws of Nature*. New York: Random House.

On the theoretical virtues: Schindler, S. (2018). *Theoretical Virtues in Science: Uncovering Reality through Theory*. Cambridge: Cambridge University Press.

On the rules of scientific debate: Strevens, M. (2020). *The Knowledge Machine: How Irrationality Created Modern Science*. New York: W.W. Norton.

On the role of theoretical virtues in the success of science: Trout, J. D. (2016). *Wondrous Truths: The Improbable Triumph of Modern Science*. Oxford: Oxford University Press.

On young children's preference for simpler theories: Walker, C. M., Williams, J. J., Lombrozo, T., Rafferty, A. N., and Gopnik, A. (2017). Explaining constrains causal learning in childhood. *Child Development* 88: 229–246.

Chapter 6

On scientific understanding: de Regt, H. (2017). *Understanding Scientific Understanding*. Oxford: Oxford University Press; Elgin, C. Z. (2017). *True Enough*. Cambridge, MA: MIT Press; Kampourakis, K. and McCain, K. (2019). *Uncertainty: How It Makes Science Advance*. New York: Oxford University Press.

On the prevalence of cognitive biases: Gilovich, T. (1991). *How We Know What Isn't So: The Fallibility of Human Reason in Everyday Life*. New York: Free Press.

On the possibility that studies purporting to show cognitive errors are mistaken: Hertwig, R. and Gigerenzer, G. (1999). The "conjunction fallacy" revisited: How intelligent inferences look like reasoning errors. *Journal of Behavioral Decision Making* 12: 275–305; Mousavi, S. and Gigerenzer, G. (2011). Revisiting the "error" in studies of cognitive errors. In D. A. Hofmann and M. Frese (Eds.). *Error in Organizations*. New York: Taylor & Francis, 97–112.

On what makes an explanation the best: Lipton, P. (2004). *Inference to the Best Explanation*, 2nd ed. New York: Routledge.

On the inherent uncertainty of science: Rovelli, C. (2014). Science is not about certainty. In J. Brockman (Ed.). *The Universe: Leading Scientists Explore the Origin, Mysteries, and Future of the Cosmos*. New York: Harper Perennial, 214–228; Kampourakis, K. and McCain, K. (2019). *Uncertainty: How It Makes Science Advance*. New York: Oxford University Press.

On reasons to be skeptical of felt understanding: Trout, J. D. (2016). *Wondrous Truths: The Improbable Triumph of Modern Science*. Oxford: Oxford University Press.

Chapter 7

On understanding generated from idealizations: de Regt, H. (2017). *Understanding Scientific Understanding*. Oxford: Oxford University Press; Elgin, C. Z. (2017). *True Enough*. Cambridge, MA: MIT Press.

On the nature of scientific models: Giere, R. N. (2010). *Scientific Perspectivism*. Chicago, IL: University of Chicago Press.

On uncertainty resulting from idealizations: Kampourakis, K. and McCain, K. (2019). *Uncertainty: How It Makes Science Advance*. New York: Oxford University Press; Rovelli, C. (2014). Science is not about certainty. In J. Brockman (Ed.), *The Universe: Leading Scientists Explore the Origin, Mysteries, and Future of the Cosmos*. New York: Harper Perennial, 214–228.

On idealizations in biology: Sober, E. (2000). *Philosophy of Biology*, 2nd ed. Boulder, CO: Westview Press.

On the nature of idealizations: Strevens, M. (2008). *Depth: An Account of Scientific Explanation*. Cambridge, MA: Harvard University Press.

On the inescapability of idealizing models in science: Potochnik, A. (2017). *Idealization and the Aims of Science*. Chicago, IL: University of Chicago Press; Trout, J. D. (2016). *Wondrous Truths: The Improbable Triumph of Modern Science*. Oxford: Oxford University Press.

Chapter 8

On possible explanations for the origin of life on Earth: Bernhardt, H. S. (2012). The RNA world hypothesis: The worst theory of the early evolution of life (except for all of the others). *Biology Direct* 7: 23.

On the idea that knowledge only comes by inference to the only explanation: Bird, A. (2022). *Knowing Science*. Oxford: Oxford University Press.

On the requirement that inferred explanations must not only be the best, but also "good enough": Dellsén, F. (2021). Explanatory consolidation: From "best" to "good enough". *Philosophy and Phenomenological Research*; Lipton, P. (2004). *Inference to the Best Explanation*, 2nd ed. New York: Routledge; McCain, K. and Poston, T. (2019). Dispelling the disjunction objection to explanatory inference. *Philosophers' Imprint* 19: 1–8.

On successes of IBE in science: Douven, I. (2017). Abduction. *The Stanford Encyclopedia of Philosophy* (Summer 2017 Edition), Edward N. Zalta (Ed.). Available at: https://plato.stanford.edu/archives/sum2017/entries/abduction; T rout, J. D. (2016). *Wondrous Truths: The Improbable Triumph of Modern Science*. Oxford: Oxford University Press.

On the relation of scientific and everyday thinking: Einstein, A. (1936). Physics and reality. *Journal of the Franklin Institute* 221: 349–382.

On the explanatory power of the double-helix structure of DNA: Kampourakis, K. (2017). *Making Sense of Genes*. Cambridge: Cambridge University Press.

On how scientific explanations lead to scientific knowledge: Kampourakis, K. and McCain, K. (2019). *Uncertainty: How It Makes Science Advance*. New York: Oxford University Press; McCain, K. (2019). How do explanations lead to scientific knowledge? In K. McCain and K. Kampourakis (Eds.). *What Is Scientific Knowledge? An Introduction to Contemporary Epistemology of Science*. New York: Routledge, 52–65; McCain, K. (2020). What is biological knowledge? In K. Kampourakis and T. Uller (Eds.). *Philosophy of Science for Biologists*. Cambridge: Cambridge University Press, 36–54.

On how the changes in our scientific practices make our best hypotheses likely to be true: Psillos, S. (1999). *Scientific Realism: How Science Tracks Truth*. New York: Routledge.

On the Best of a Bad Lot and unconceived rival hypotheses objections to IBE: van Fraassen, B. (1989). *Laws and Symmetry*. Oxford: Oxford University Press.

Figure Credits

Figure 4.1 Reproduced from Sahin, O., Salim, H., Suprun, E., et al. (2020). Developing a preliminary causal loop diagram for understanding the wicked complexity of the COVID-19 pandemic. Systems 8:20. https://doi.org/10.3390/systems8020020 Open Access.

Figure 4.2 Reproduced from UK Government Office for Science. https://assets.publishing.service.gov.uk/government/uploads/system/uploads/attachment_data/file/296290/obesity-map-full-hi-res.pdf Licensed under Open Government License v3.0.

Index